Essential Maple

Dept.
Applied Maths.
U W O

Robert M. Corless

Essential Maple

An Introduction for Scientific Programmers

With 32 Figures

Springer-Verlag
New York Berlin Heidelberg London Paris
Tokyo Hong Kong Barcelona Budapest

Robert M. Corless
Department of Applied Mathematics
University of Western Ontario
London, Ontario N6A 5B7
Canada

Cover illustration: The cover shows the graph of the Gauss Map $G(x) = 1/x$ mod 1 displayed as a torus. See Chapter 2 for more details.

The production staff at Springer-Verlag has included in the design of this book the overflow of mathematical formulas into the margins. This emphasizes the formulas produced by the Maple computer algebra language while maintaining a readable, visually pleasing text layout.

Library of Congress Cataloging-in-Publication Data

Corless, Robert M.
    Essential Maple: an introduction for scientific programmers /
  Robert M. Corless.
        p.   cm.
    Includes bibliographical references and index.
    ISBN 0-387-94209-2.
    1. Maple (Computer file) 2. Mathematics—Data processing.
  I. Title.
  QA76.95.C68 1995
  510'.285'53—dc20   94-35231

Printed on acid-free paper.

Production managed by Natalie Johnson; manufacturing supervised by Jacqui Ashri.
Typeset by Digital Graphics, Inc., using the author's LATEX files.
Printed and bound by Hamilton Printing Co., Castleton-on-Hudson, NY.
Printed in the United States of America.

9 8 7 6 5 4 3 2 1

ISBN 0-387-94209-2 Springer-Verlag New York Berlin Heidelberg
ISBN 3-540-94209-2 Springer-Verlag Berlin Heidelberg New York

For my parents:

John D. Corless and Marion L. Corless
and
M. Aly Hassan and Galima Hassan

Preface

What's in this book

This book contains an accelerated introduction to Maple, a computer algebra language. It is intended for scientific programmers who have experience with other computer languages such as C, FORTRAN, or Pascal. If you wish a longer and more leisurely introduction to Maple, see (8, 27, 39).

This book is also intended as a reference summary for people who use Maple infrequently enough so that they forget key commands. Chapter 4 is a keyword summary. This will be useful if you have forgotten the exact Maple command for what you want. This chapter is best accessed through the table of contents, since it is organized by subject matter.

The mathematical prerequisites are calculus, linear algebra, and some differential equations. A course in numerical analysis will also help. Any extra mathematics needed will be developed in the book.

This book was prepared using Maple V Release 3, although most of the examples will work with, at most, only slight modification in Maple V Release 2. This book does not require any particular hardware. The systems I have used in developing the book are machines running IBM DOS and WIN/OS2, Unix machines in an ASCII terminal mode, and X-windows systems. There should be no adjustments necessary for readers equipped with Macintoshes or other hardware.

Maple is an evolving system. New features will be described in the documentation for updates (?updates in Maple).

How to read this book

The suggested way to read this book is to read sections 1.1–1.5 at a sitting, while you have Maple running in front of you so you can try things out. Read the rest of the book at your leisure, and in any order you like. I suggest that you look at the keyword summary (Chapter 4) and use that as a guide to further research. The table of contents provides the best index into the keyword summary. Check ?updates if you are using a version later than Release 3.

The exercises are intended to give you practice in what has just been shown and to develop the ideas further. They vary in difficulty from trivial to quite difficult. They have been used as assignments in an introductory graduate course in applied computer algebra here at the University of Western Ontario. It is not necessary to do them to benefit from this book, but it's probably more fun than just reading.

There are many small programs scattered throughout this book, and I hope that you may find them useful in themselves, and as guides for writing your own.

Acknowledgements

The most significant help I received for this book is from my wife, Sumaya. By hard work in a wide variety of capacities, she has made writing this book both possible and very pleasant. She deserves much more credit than she gets from this one little paragraph.

My daughter Shamila, on the other hand, hasn't really been helpful at all—but she always *wants* to help, and somehow that's just as good, from someone who's five years old.

My parents (both sets), to whom this book is dedicated, were invaluable.

For technical help, thanks go to Keith Geddes for getting me interested in computer algebra with the first Maple course offered at the University of Waterloo. The course began with ALTRAN and finished with Maple (this was back before version numbers). I have since used computer algebra in nearly all my work, both research and teaching. The other major influence on my computer algebra career is David Jeffrey, who taught me what it means to do research in applied mathematics and has continued as a good friend and collaborator. The members of the watmaple mailing group, past and present, from Gaston Gonnet and Michael Monagan through to the most recent student research assistant, have participated in many extremely interesting discussions and have taught me a lot about Maple. George Corliss, Dave Hare, Henning Rasmussen, and Kelly Roach provided particularly detailed criticisms of early drafts of this book. Bill Bauldry and an anonymous reviewer also provided helpful remarks. Niklaus Mannhart helped with some final LATEX work, as well as reading over the manuscript.

Portions of this book were finalized while I was on sabbatical at T. J. Watson Research Center in Yorktown Heights, New York. Stephen

Watt, Dick Jenks, and Tim Daly were generous with their time and energy, even while on a tight schedule.

Thanks also go to Darren Redfern for excellence as an "author's editor"; to Stan Devitt for developing the processing tools used to efficiently include Maple input and output in this LATEX document; to my students Mohammed O. Ahmed, Anne-Marie E. Allison, Tianhong Chen, and David W. Linder for being 'lab animals' in testing this book out; and similarly to all my Applied Computer Algebra students, past and present, for helping me refine my ideas and presentations.

Contents

List of Figures

Basics

1

'But the mad will ne'er content, till he shall have patterned out
to his own most mathematical likings the unpeerable
inventions of God...'
—E. R. Eddison, *A Fish Dinner in Memison*, p. 146.

This chapter shows how to get started in Maple, gives some sample sessions, discusses some common difficulties and errors, and lays a firm groundwork for more advanced use of Maple. Most important are sections 1.1 to 1.5—the others can be read after you skim the rest of the book.

1.1   Getting started

To start Maple on most computer systems, type `maple` (or `xmaple` if you wish to start an X-Windows Maple session), or click on the Maple icon. Consult your local wizard if this doesn't work. To get help once you have started Maple, type `?` after the Maple prompt, which is usually an angle bracket (>). To get help on a particular Maple topic, type `?<keyword>` where `<keyword>` is the Maple word for what you want help on. For example, this is part of what you get if you type `?index` after the prompt.

```
> ?index
    HELP FOR: Index of help descriptions

    CALLING SEQUENCE:

    ?index[<category>]    or    help(index, <category>);

    SYNOPSIS: ...
```

If you don't know the exact word, look in the keyword summary in Chapter 4 of this book, or simply try a few alternatives: Maple will try to help you locate what you want. If your system has a help browser, use that.

Exercises

1. Find out how to use the linear algebra package by starting Maple and issuing the commands ?linalg and ?with. Explore at least one routine (e.g., jordan).

2. Find out how to use the student package by issuing the command ?student. Explore at least one routine (e.g., changevar).

1.1.1   Basic command syntax

Note that Maple is *case-sensitive*, so series is different from SERIES is different from Series. The Maple command is series.

```
> series(sin(exp(x)-1), x);
```

$$x + \frac{1}{2} x^2 - \frac{5}{24} x^4 - \frac{23}{120} x^5 + O(x^6)$$

That is one way to compute a series in Maple. But if instead we use some uppercase letters, Maple thinks we're talking about some other function, that it may learn about later.

```
> Series(sin(x), x);
```

$$\text{Series}(\sin(x), x)$$

As you see, Maple echoes syntactically legal input that it doesn't understand. This behaviour is fundamental to Maple's ability as a symbolic processor, but in this case it may not be what is wanted. In particular, if you have the 'caps lock' key on, you may get something like the following.

```
> SERIES(cos(x), x);
```

$$\text{SERIES}(\cos(x), x)$$

Maple statements end with a semicolon (;) or colon (:). Statements ending in a colon (:) perform computations but the results are not printed. This is used to suppress the printing of voluminous intermediate results. For example,

```
> expand( (x+y)^3 );
```

$$x^3 + 3 x^2 y + 3 x y^2 + y^3$$

displays its results, as expected. On the other hand, suppose we wish to compute the coefficient of x^{48} in $(x + 3)^{100}$. Then the intermediate result below is not of any real interest, and since it occupies 262 lines it is a good idea not to print it.

```
> expand( (x+3)^100 ):
> coeff(", x, 48);
```

6022152095047244121129504670562748296126762437703 45100

The ditto (") refers to the previous result.

A *common mistake*: if you forget to type the statement terminator (a colon or semicolon), *do not* retype the line—simply enter a colon or semicolon. If you re-type the line you will (probably) introduce a syntax error as Maple will try to interpret what you have typed twice as a single, multi-line, Maple statement. Other common mistakes are covered in section 1.1.4.

```
> expand( (x+3)^100 )
> :
> coeff(", x, 48)
> ;
```

6022152095047244121129504670562748296126762437703 45100

1.1.2 Sample Maple sessions

Three short Maple sessions follow. You should start Maple up on your system, and type in the following commands.

1.1.2.1 First sample session—Maple as calculator

This session shows how to use Maple to solve some problems in algebra, linear algebra, and calculus. We begin by factoring a polynomial.

```
> factor(t^12 - 1);
```

The output from this command is shown below. It is possible that the ordering of the factors may be different in your session.

$$(t-1)(t+1)(t^2+t+1)(t^2-t+1)(t^2+1)(t^4-t^2+1)$$

Is that factorization correct? We can see by inspection that the roots ± 1 are included in those factors, as are the roots $\pm i$ (where i is the square root of -1). So we tend to believe that Maple got that factorization right, and of course we can ask Maple to expand that factorization out to get back $t^{12} - 1$.

```
> expand(");
```

$$t^{12} - 1$$

Now let us do some simple computations from linear algebra.

```
> with(linalg):
Warning: new definition for    norm
Warning: new definition for    trace
```

The call with(linalg) loads the linear algebra package into Maple. This enables simple access to Maple's linear algebra package. The warnings tell you that the main library routines norm and trace have been replaced by linear algebra routines with the same names. Ignore these warnings here, because we were not intending to use those main routines. All future occurrences of this message will be suppressed in this text.

```
> A := matrix([[4,5], [5,6]]);
```

$$A := \begin{bmatrix} 4 & 5 \\ 5 & 6 \end{bmatrix}$$

Now compute the characteristic polynomial of that matrix.

```
> p := charpoly(A, lambda);
```

$$p := \lambda^2 - 10\,\lambda - 1$$

The trace of the matrix is $6 + 4 = 10$, which should be the negative of the linear coefficient. It is. The determinant of the matrix is $4 \cdot 6 - 5^2 = -1$, which should be the constant coefficient. Again, it is. As before we conclude that Maple got it right.

Now let us do some calculus.

```
> Int( 1/(t^6 - 1), t) = int( 1/(t^6-1), t);
```

The output from this command appears below. Let us first examine the input. Note that the left-hand side of the input is the same as the right-hand side, except that the left-hand side has a capitalized Int (which makes the command *inert*. We will discuss inert functions in section 1.8.1). Here it is used to produce a sensible equation, with an integral on the left-hand side of the output below, put equal to an expression on the right-hand side.

$$\int \frac{1}{t^6-1}\,dt = \frac{1}{6}\ln(t-1) - \frac{1}{12}\ln(t^2+t+1) - \frac{1}{6}\sqrt{3}\arctan\left(\frac{1}{3}(2t+1)\sqrt{3}\right)$$
$$- \frac{1}{6}\ln(t+1) + \frac{1}{12}\ln(t^2-t+1) - \frac{1}{6}\sqrt{3}\arctan\left(z\frac{1}{3}(2t-1)\sqrt{\ }\right.$$

Note that Maple did not add an arbitrary constant to its answer. This is supposed to be understood, and if you wish to have the constant there explicitly, you must put it in yourself by adding it on, as in `int(1/(t^6-1), t) + C;`.

The returned answer above looks formidable. If we wish to check that answer independently from Maple, we would most likely prefer to do it numerically. However, the code in Maple for differentiation is independent of the code for integration, so if we ask Maple to differentiate both sides of the above equation we will get a useful confirmation.

```
> diff(", t);
```

$$\frac{1}{t^6-1} = \frac{1}{6}\frac{1}{t-1} - \frac{1}{12}\frac{2t+1}{t^2+t+1} - \frac{1}{3}\frac{1}{1+\frac{1}{3}(2t+1)^2}$$
$$- \frac{1}{6}\frac{1}{t+1} + \frac{1}{12}\frac{2t-1}{t^2-t+1} - \frac{1}{3}\frac{1}{1+\frac{1}{3}(2t-1)^2}$$

Both sides were differentiated—now we have only to simplify the results. It turns out, after some experimentation, that the command that simplifies things most efficiently is `normal` with the `expanded` option.

```
> normal(", expanded);
```

$$\frac{1}{t^6-1} = \frac{1}{t^6-1}$$

So it appears that Maple found a correct antiderivative for $1/(t^6-1)$. Now let us solve the logistic differential equation.

```
> logistic := diff(x(t), t) = x(t)*(1-x(t));
```

$$\text{logistic} := \frac{\partial}{\partial t}x(t) = x(t)(1-x(t))$$

Let us use the initial condition $x(0) = \alpha$.

```
> initial_cond := x(0) = alpha;
```

$$\text{initial\_cond} := x(0) = \alpha$$

Now solve the differential equation with this initial condition for the un-known $x(t)$.

```
> ans := dsolve( {logistic, initial_cond}, x(t) );
```

$$ans := x(t) = -\frac{1}{-1 - \dfrac{e^{-t}(1-\alpha)}{\alpha}}$$

We check the result by substituting it back into the differential equation, and verifying the initial condition.

```
> subs(ans, logistic);
```

$$\frac{\partial}{\partial t}\left(-\frac{1}{-1 - \dfrac{e^{-t}(1-\alpha)}{\alpha}}\right) = -\frac{1 + \dfrac{1}{-1 - \dfrac{e^{-t}(1-\alpha)}{\alpha}}}{-1 - \dfrac{e^{-t}(1-\alpha)}{\alpha}}$$

```
> normal(");
```

$$-\frac{\alpha e^{-t}(-1+\alpha)}{(-\alpha - e^{-t} + e^{-t}\alpha)^2} = -\frac{\alpha e^{-t}(-1+\alpha)}{(-\alpha - e^{-t} + e^{-t}\alpha)^2}$$

Since the left- and right-hand sides are equal, $x(t)$ is indeed a solution.

```
> subs(t=0, ans);
```

$$x(0) = -\frac{1}{-1 - \dfrac{e^0(1-\alpha)}{\alpha}}$$

```
> normal(");
```

$$x(0) = \alpha$$

And the initial condition is satisfied.

1.1.2.2 Terminating a Maple session

You quit Maple by issuing the commands quit, done, or stop, or by choosing Exit from the menu, if you are using a menu-driven system. These statements can be terminated by a carriage return or enter—no colon or semicolon is necessary. Now quit the Maple sample session.

```
> quit
```

This ends the first sample session. You can get help on the meaning of each of the commands used above by the **?** command, as noted previously.

Exercises

1. Factor $t^{24} - 1$. Check your answer.
2. Factor $t^6 - 1$ down to linear factors. Hint: see `?factor`, and you need the extension $K = \sqrt{-3}$. Again (and always) check your answer.
3. Find the inverse of the matrix A from this sample session, using Maple (see `?inverse` or `?evalm`).
4. Find $\int \exp(-t)\sin(t)\, dt$.
5. Find the fifth derivative of $\exp(\theta)\sin^{-1}\theta$. (The inverse sine function $\sin^{-1}$ is called `arcsin` in Maple.)
6. Solve $x^2 y'' + xy' + y = 3x^3$.

1.1.2.3   Second session—'Hello, world'

For the second sample session, we write the obligatory 'Hello, world' program (or a slight variant of it). Start Maple again, and enter the following one-line program.

```
> hi := proc() 'Hello, Worf' end;
```

The quotes are left quotes (`) and not right quotes (').

```
        hi := proc() 'Hello, Worf' end
```

```
> hi(Maple);
```

$$\text{Hello, Worf}$$

Now quit Maple again.

```
> quit
```

A Maple procedure body is begun with a `proc()` keyword and ended with the end keyword. It takes arguments, which may or may not be indicated in the `proc()` keyword. To give a name to a procedure, you assign the procedure body to the name. In the above example, the name of the procedure was `hi`.

Exercises

1. Compare the results of the following Maple input statements (`hi` is defined as above).

   ```
   > hi;
   > hi();
   > hi(Ginger);
   > eval(hi);
   ```

2. What will the following program do?

   ```
   > hi := proc(x)
   >    if x=Maple then
   >       'Hello, Deanna'
   >    else
   >       'I beg your pardon?'
   >    fi
   > end:
   ```

1.1.2.4   Third session—heat conduction by Fourier series

For our final sample session, we attempt something a little more ambitious, namely the solution of the heat equation $u_t = u_{xx}$ on $0 \le x \le 1, 0 < t$, with boundary conditions $u(0,t) = u(1,t) = 0$, and initial condition $u(x,0) = x^2(1 - x^2)$. This will give us the nondimensional temperature u for all later times t in a rod of length 1 whose initial temperature is distributed as $x^2(1 - x^2)$.

We use the standard theory of Fourier series to solve this problem (see, for example, (6)), and use Maple as a worksheet for the calculations. This gives a quick overview of integration, some algebraic manipulation facilities, and some elementary plotting features. However, it does use a few advanced features of Maple. These may appear to be somewhat mysterious at this stage. I urge you to follow along through the sample session as far as you can, and skim until the end of the session if you get into trouble and can't get out—you can always return to this example later. As this session proceeds, issue help commands as needed (for example, `?int`).

The Fourier series solution we are looking for is

$$u(x,t) = \sum_{k=1}^{\infty} c_k e^{-k^2\pi^2 t} \sin(k\pi x) \, ,$$

and we will use Maple below to calculate the constants c_k. It is known that

$$c_k = \frac{\int_0^1 f(x) \sin(k\pi x)\, dx}{\int_0^1 \sin^2(k\pi x)\, dx}$$

so we begin with the evaluation of these integrals.

```
> int( sin(k*Pi*x)^2, x=0..1);
```

$$\frac{-\cos(\,k\,\pi\,)\sin(\,k\,\pi\,) + k\,\pi}{2\,k\,\pi}$$

It seems obvious to us that the above expression can be simplified, but we must remember that Maple does not know that k is an integer. The simplest way to help Maple out is to use our own knowledge.

```
> subs(sin(k*Pi)=0, ");
```

$$\frac{1}{2}$$

This gives us one of the integrals in the definition of c_k. Now for the other.

```
> int( x^2*(1-x^2)*sin(k*Pi*x), x=0..1);
```

$$-2\,\frac{-12\cos(\,k\,\pi\,) + 5\,k^2\,\pi^2\cos(\,k\,\pi\,) - 12\,k\,\pi\sin(\,k\,\pi\,) + k^3\,\pi^3\sin(\,k\,\pi\,)}{k^5\,\pi^5}$$
$$-\,2\,\frac{k^2\,\pi^2 + 12}{k^5\,\pi^5}$$

```
> subs(sin(k*Pi)=0, ");
```

$$-2\,\frac{-12\cos(\,k\,\pi\,) + 5\,k^2\,\pi^2\cos(\,k\,\pi\,)}{k^5\,\pi^5} - 2\,\frac{k^2\,\pi^2 + 12}{k^5\,\pi^5}$$

```
> subs(cos(k*Pi)=(-1)^k, ");
```

$$-2\,\frac{-12\,(-1\,)^k + 5\,k^2\,\pi^2\,(-1\,)^k}{k^5\,\pi^5} - 2\,\frac{k^2\,\pi^2 + 12}{k^5\,\pi^5}$$

```
> expand(");
```

$$24\,\frac{(-1\,)^k}{k^5\,\pi^5} - 10\,\frac{(-1\,)^k}{k^3\,\pi^3} - 2\,\frac{1}{k^3\,\pi^3} - 24\,\frac{1}{k^5\,\pi^5}$$

```
> collect(", k, factor);
```

$$-2\,\frac{5\,(-1\,)^k + 1}{\pi^3\,k^3} + 24\,\frac{(-1\,)^k - 1}{\pi^5\,k^5}$$

Now we know that c_k is the above value divided by $1/2$. The following is a convenient way to express this as a *functional operator*.

```
> c := unapply(2*", k);
```

$$c := k \rightarrow -4\,\frac{5\,(-1)^k+1}{\pi^3\,k^3}+48\,\frac{(-1)^k-1}{\pi^5\,k^5}$$

The name unapply is not evocative of the procedure's purpose until you think of *applying* an operator, say $g : k \mapsto k^3 + 5$, to an argument, say m, to get the *expression* $m^3 + 5$. The opposite to this process, that of converting an expression to an operator, is then reasonably thought of as "unapplication".

Returning to our example, we could express the sum to n terms of the Fourier series for the initial condition $f(x)$ as follows.

```
> fq := n -> Sum( c(k)*sin(k*Pi*x), k=1..n);
```

$$\text{fq} := n \rightarrow \sum_{k=1}^{n} c(k)\sin(k\pi x)$$

It turns out, however, to be better to use the more efficient (though uglier) construct given below. See the exercises below, and the incidental discussion of seq in section 2.7.2.

```
> fn := n -> convert([seq(c(k)*sin(k*Pi*x), k=1..n)], '+');
Warning, 'k' is implicitly declared local
```

$$\text{fn} := n \rightarrow \text{local } k; \text{convert}([\,\text{seq}(c(k)\sin(k\pi x), k=1..n)\,], `+`)$$

Ignore that warning. It is telling us that the k in the sum is independent of any other k that we may have been using. I cannot foresee any circumstances where this would not be what is wanted. All future occurrences of this message will be suppressed in this text. Now we return to the example. Let us take five terms in the series and investigate the error.

```
> f5 := fn(5);
```

$$f5 := \left(16\frac{1}{\pi^3}-96\frac{1}{\pi^5}\right)\sin(\pi x)-3\frac{\sin(2\pi x)}{\pi^3}+\left(\frac{16}{27}\frac{1}{\pi^3}-\frac{32}{81}\frac{1}{\pi^5}\right)\sin(3\pi x)$$
$$-\frac{3}{8}\frac{\sin(4\pi x)}{\pi^3}+\left(\frac{16}{125}\frac{1}{\pi^3}-\frac{96}{3125}\frac{1}{\pi^5}\right)\sin(5\pi x)$$

```
> plot(f5 - x^2*(1-x^2), x=0..1);
```

See Figure 1.1. The maximum magnitude of the error is less than 0.01, with only five terms in the series. The maximum value of the function occurs

Figure 1.1 Graph of the error in the five-term solution to the heat equation problem

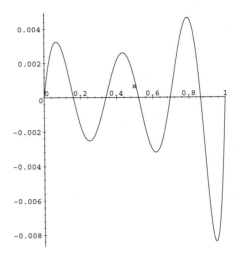

when $x^2 = 1 - x^2$ (by the Arithmetic-Geometric Mean Inequality (37), or by elementary calculus), or $x^2 = 1/2$ so $f(x) = 1/4$ at this point. Thus the relative error is less than 5%. Let us see what happens if we take ten terms. Here we don't want to *look* at the ten terms in the series, just compute them.

```
> f10 := fn(10):
> plot(f10 - x^2*(1-x^2), x=0..1);
```

Figure 1.2 Graph of the error in the ten-term solution to the heat equation problem

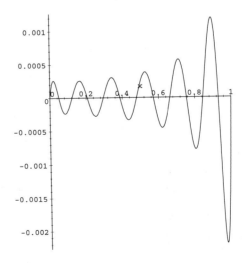

See Figure 1.2. The maximum magnitude of the error now is roughly 0.002, or less than 1%.

Now let us consider the solution u—does it satisfy the differential equation? The Fourier series was constructed so that each term *should* satisfy the differential equation, so the only error we are expecting to commit is the representation of the initial condition. We check that our solution satisfies the differential equation to guard against blunders, as opposed to approximation errors.

```
> un := n -> convert( [seq(c(k)*exp(-k^2*Pi^2*t)*sin(k*Pi*x), k=1..n)],
```

$$un := n \to \text{local } k; \text{ convert} \left(\left[\text{seq} \left(c(k) e^{(-k^2 \pi^2 t)} \sin(k \pi x), k = 1..n \right) \right], \text{`}+\text{`} \right)$$

```
> u5 := un(5);
```

$$u5 := \left(16 \frac{1}{\pi^3} - 96 \frac{1}{\pi^5} \right) e^{(-\pi^2 t)} \sin(\pi x) - 3 \frac{e^{(-4\pi^2 t)} \sin(2\pi x)}{\pi^3}$$

$$+ \left(\frac{16}{27} \frac{1}{\pi^3} - \frac{32}{81} \frac{1}{\pi^5} \right) e^{(-9\pi^2 t)} \sin(3\pi x) - \frac{3}{8} \frac{e^{(-16\pi^2 t)} \sin(4\pi x)}{\pi^3}$$

$$+ \left(\frac{16}{125} \frac{1}{\pi^3} - \frac{96}{3125} \frac{1}{\pi^5} \right) e^{(-25\pi^2 t)} \sin(5\pi x)$$

```
> diff(u5, t) - diff(u5, x, x);
```

$$0$$

```
> u10 := un(10):
> diff(u10, t) - diff(u10, x, x);
```

$$0$$

Evidently it does satisfy the differential equation.

We should also check that the boundary conditions $u(0, t) = u(1, t) = 0$ are satisfied, but this is obvious from the fact that each term is multiplied by $\sin(k\pi x)$ for some k.

Now let us draw a contour plot of this function of x and t. After some experimentation, we find that the following scale gives useful information.

```
> plots[contourplot](u10, x=0..1, t=0..0.2, grid=[30,30], colour=black);
```

Figure 1.3 Lines of equal temperature in the x-t plane

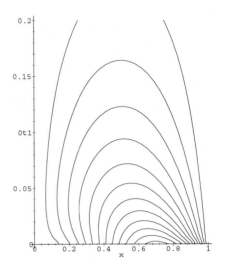

That plot, shown in Figure 1.3, shows *isotherms*, or lines of equal temperature. We see that for small times and x less than about 0.4, the temperature
rises initially and then falls. This makes sense, as the initially peaked temperature distribution (which has its maximum at $x = 1/\sqrt{2}$) is smoothed
out as time progresses and the initially colder parts of the rod are warmed
by the hotter adjacent region.

Now suppose we wished to repeat this process for several different
initial functions $f(x)$. It would make sense to write a Maple program to
automate those steps. This is probably the main strength of Maple, or
indeed of any computer algebra language—it is a high-level programming
language, which you can customize to your own needs.

```
#
# PROGRAM:        Fourier_sine:  compute the Fourier sine series of an
#                                input function.
# MAINTENANCE HISTORY:
#    First version                    Feb 17 1994.
#    Modified to increase readability   May 25 1994.
# BASIC IDEA:     Standard theory of Fourier series.
# REFERENCE:      William E. Boyce and Richard C. DiPrima,
#                 Elementary Differential Equations and
#                 Boundary Value Problems, 2nd. ed., Wiley, 1969,
#                 pp. 423--429.

# CALLING SEQUENCE: Fourier_sine(fn);
# INPUT:  fn : an expression denoting a function of x, e.g.,
#              sin(x) or cos(x^3).

# OUTPUT: an  O P E R A T O R  which takes an integer n as
#         argument and returns a sum of n terms of the
#         Fourier sine series for fn.
```

```
# SIDE EFFECTS: assigns an operator to the global variable
#               ck, where ck(m) gives the m-th coefficient
#               in the Fourier sine series.
# GLOBAL VARIABLES:  x   and   ck.
#
#     (using the global variable x is a particularly bad
#      programming practice, but is used here for simplicity
#      because this program is not intended for re-use.  If
#      it ever DOES get re-used, the first change made will
#      be to add the variable of the function to the
#      calling sequence: Fourier_sine(sin(t), t);    )
# KNOWN BUGS/WEAKNESSES:   May occasionally divide by zero.
#                          See the exercises.  The 'has'
#                          test for integrals, to see if
#                          the integral succeeded, may also
#                          fail, or be expensive.  The new
#                          routine 'hasfun' (for the next
#                          release of Maple) is intended to
#                          replace this.
# Miscellaneous Remarks:   Using global variables generally
#                          conflicts with other people's programs.
#                          This program is not re-usable.
Fourier_sine := proc(f)
    local k; global x, ck;

    ck := 2*int( f*sin(k*Pi*x), x=0..1);

    if has(ck, int) then
        ERROR('Sorry, couldn't do the integral')
    else
        ck := subs(sin(k*Pi)=0, cos(k*Pi)=(-1)^k, ck);
        ck := unapply(collect(ck, k, factor), k);
        n -> convert( [seq(ck(k)*sin(k*Pi*x), k=1..n)], '+')
    fi
end:
```

The procedure has a side effect: it defines the global variable ck as an *operator* that gives the coefficients.

Now we test it.

```
> Fourier_sine( x^2*(1-x^2) );
```

$$n \rightarrow \text{local } k; \text{convert}([\text{seq}(\text{ck}(k) \sin(k \pi x), k = 1..n)], ' + ')$$

```
> G2 := ":
> G2(5);
```

$$\left(16 \frac{1}{\pi^3} - 96 \frac{1}{\pi^5}\right) \sin(\pi x) - 3 \frac{\sin(2 \pi x)}{\pi^3} + \left(\frac{16}{27} \frac{1}{\pi^3} - \frac{32}{81} \frac{1}{\pi^5}\right) \sin(3 \pi x)$$
$$- \frac{3}{8} \frac{\sin(4 \pi x)}{\pi^3} + \left(\frac{16}{125} \frac{1}{\pi^3} - \frac{96}{3125} \frac{1}{\pi^5}\right) \sin(5 \pi x)$$

```
> "-f5;
```

$$0$$

So we get the same answer as we did before. Let us try a different function.

```
> C := Fourier_sine( cos(x) );
```

$$C := n \rightarrow \text{local } k; \text{convert}([\, \text{seq}(\text{ck}(k) \sin(k \, \pi \, x), k = 1..n)\,], ` + `)$$

```
> C(5);
```

$$-\frac{(-2\cos(1)\pi - 2\pi)\sin(\pi x)}{(1+\pi)(-1+\pi)} - \frac{(4\cos(1)\pi - 4\pi)\sin(2\pi x)}{(1+2\pi)(-1+2\pi)}$$
$$-\frac{(-6\cos(1)\pi - 6\pi)\sin(3\pi x)}{(1+3\pi)(-1+3\pi)} - \frac{(8\cos(1)\pi - 8\pi)\sin(4\pi x)}{(1+4\pi)(-1+4\pi)}$$
$$-\frac{(-10\cos(1)\pi - 10\pi)\sin(5\pi x)}{(1+5\pi)(-1+5\pi)}$$

```
> plot("-cos(x), x=0..1);
```

Figure 1.4 Graph of the error in the five-term solution with a nonsmooth initial condition

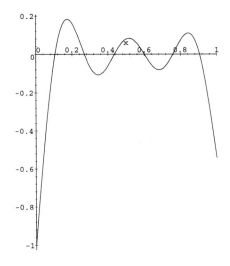

That plot is shown in Figure 1.4. We see the Gibbs phenomenon—the overshoot near the discontinuity (6)—clearly in that plot. The convergence is less good for this example because the odd extension of $\cos(x)$ to $-1 \le x \le 1$ has a jump discontinuity at the origin.

The program `Fourier_sine` is not intended as an example of elegant, robust programming. It violates several 'good programming style' rules—for example, it has a side effect: namely, it changes the value of the global variable ck. For another, it uses the global variable x on input. However, the program is *not* intended for general-purpose use: it is an example of a 'throwaway' program written quickly to solve one particular problem. I find Maple extremely useful in a wide variety of problems, and the creation of such 'throwaway' programs to be a major help in day-to-day computation.

It is also easy to write much more powerful, robust, and general-purpose programs in the Maple programming language. For examples of such programs, you can look at the source code for the Maple library itself, or in (12), or in (20). The present book limits its scope to teaching the use of Maple as a calculator and the construction of small to moderately small programs.

You may wish to save your work before quitting.

```
> save Fourier_sine, 'myfile.mpl':
> quit
```

This creates a human-and-Maple-readable file called `myfile.mpl` containing `Fourier_sine`. Note that it doesn't save the comments: that can be done in a worksheet, but the comments above were imported from a file in which `Fourier_sine` was defined. I used an editor to create the program, and read it in.

Exercises

1. After `Fourier_sine(x^2*(1-x^2))` has been called, what does ck(m) give you for a variable m?

2. Solve $u_t = u_{xx}$ on $0 \leq x \leq 1$ for $t > 0$ given that $u(x, 0) = x^3(1 - x^3)$, $u(0, t) = u(1, t) = 0$.

3. If instead $u(x, 0) = x \sin \pi x$, the program gives a division by zero when computing, say, f10. Why? What can be done to solve the problem in this case? What about $u(x, 0) = x \sin^2 \pi x$?

4. Rerun the worksheet using

```
> fn := n -> Sum( c(k)*sin(Pi*x), k=1..n);
```

instead of the ugly but efficient seq construct. Comment on the execution speed difference.

1.1.3 Arithmetic

Maple has facilities for arbitrary-precision integer and rational arithmetic (both real and complex); modular arithmetic; hardware floating-point and arbitrary precision floating-point arithmetic; and the arithmetic of matrices.

The standard example of arbitrary-precision integer arithmetic is the computation of $3!!! = 6!! = 720!$, as follows. [It is the standard example because it fits nicely on the screen of a typical terminal.]

```
> 3!!!;
```

260121894356579510020490322708104361119152187501694578572754183780\
508356311569473822406785779581304570826199205758922472595366641565\
162052015873791984587740832529105244690388811884123764341191951040\
550534665861624327194019711390984553672727853709934562985558671930\
697740700037004307837589974206767840169672078462806292290321071610\
669867260548988445514257193985499448939594496064045132362140265980\
619307324936977047760606768067017649166940303481996188145562519550\
925669188308255149429475965372748456246288242342652659778973774089\
466553992435928786212515967483220976029505696699927284670563747130\
753301924831358707612541268341586012944756601145542074958995256350\
430682886346310849656506827715529962567908452357025521862223581300\
016700834523443236821935793184701956510729781804354173890560727420\
804858399591972902172661229129842051606757903623233769945396419140\
751755675576953922338030568253085999774416757843528159134613403940\
604901269542028838347101363733824484506660093348484440711931292530\
769465735433737572477223018153403264717753198453734147867432704840\
579837866187032574059389242157096959946305575210632032634932092200\
738320923356309923267504401701760572026010829288042335606643089880\
871029738079757801305604957634283868305719066220529117482251053660\
977566030295740433879834715185526028053338663571391010463364197690\
097397432285994219837046979109956303389604675889865795711176566670\
003915674815311594398004362539939973120306649060132531130471902880\
984918562037666691644687911252491937544258458950003115616829743040\
641142538074897281723375955380661719801404677935614793635266265680\
333950976000000000000000000000000000000000000000000000000000000000\
000000000000000000000000000000000000000000000000000000000000000000\
00000000000000000000000000000000000000000000000000000000

The above calculation took less than a second on a 25 MHz 486 IBM PC clone. The size of the answer usually provokes laughter, followed shortly thereafter by the question "Is it right?"

Let's explore that question, without further use of Maple. Stirling's approximate formula $n! \sim \sqrt{2\pi n} n^n \exp(-n)$ gives $720! \sim 2.60091 \cdot 10^{1746}$, and this agrees with the first three digits of the above. In the printed answer above there are 26 rows with 65 digits in them, and one with 57, giving 1747 digits in all, which agrees with the magnitude of the result of Stirling's formula. Finally, we can count the number of factors of 5 in 720!, which will give us the number of trailing zeros in the answer (because there will be more than enough factors of 2 to make each 5 a 10). The number of factors of 5 is

$$\lfloor \frac{720}{5} \rfloor + \lfloor \frac{720}{5^2} \rfloor + \lfloor \frac{720}{5^3} \rfloor + \lfloor \frac{720}{5^4} \rfloor + \lfloor \frac{720}{5^5} \rfloor + \cdots$$

and since $5^5 = 3125$, this series terminates. The value of the sum is then $144 + 28 + 5 + 1 = 178$, and we see by counting that there are 178 zeros in the answer printed above. So, in all likelihood, the answer Maple gives is correct.

There is a system-dependent size limitation on Maple integers, typically that they must have roughly fewer than 525,000 digits. This limitation is rarely exceeded in practice. For the details of the data structure used to represent integers and the reason for this magic number, see the Maple Language Reference Manual (9). For an application where this size limitation had to be overcome, see (25).

Maple fractions are simply pairs of integers, kept relatively prime by automatic GCD computations. The data type `rational` includes integers and fractions.

Exercises

1. Find the exact number of possible bridge hands. [A full deck of 52 cards is dealt randomly to four people; a 'hand of bridge' is the result of any such deal. The people are seated in a definite order.]

2. How much computer memory does a FORTRAN (or C) single or double precision floating-point number require under the IEEE standard (1)?

3. Explain the following statement and infer its proper context.

    "An n by n matrix requires $O(n^2)$ storage and $O(n^3)$ operations to invert."

In particular, explain why this is not true (or at least irrelevant) in Maple. Ignore improvements to Gaussian elimination such as Strassen's algorithm—the reason this is not true in Maple isn't because Maple can do *better*, but rather the opposite. Be concise but clear.

4. Use Newton's iteration $x_{n+1} = x_n - f(x_n)/f'(x_n)$ to generate five rational approximations to the root x^* near $x_0 = -1$ of

$$f(x) = x^3 - \frac{1}{10}x + 1 = 0 .$$

Since the error $e_n = x_n - x^*$ behaves approximately as $e_{n+1} \propto e_n^2$, estimate the error in your *best* approximation.

5. Find a cheap way to use Maple to evaluate

$$C_3^{10^{30}} \cdot \left(\frac{1}{10^{24}}\right)^3 \left(1 - \frac{1}{10^{24}}\right)^{10^{30}-3}$$

to 60 places. Note: $C_m^n = n!/(m!(n-m)!)$ is the binomial coefficient. This number is the probability that exactly three atoms out of a sample of 10^{30} decay in a period where the probability of any single atom's decay is 10^{-24}.

6. The Maple convert procedure has many uses. Convert 55/89 into 'continued fraction form'—e.g.,

$$\frac{339}{284} = 1 + \cfrac{1}{5 + \cfrac{1}{6+\frac{1}{9}}}$$

1.1.4 Common syntax errors

Like all computer languages, Maple demands a certain amount of precision from its users. This is sometimes annoying, since no one likes to get syntax error messages. However, these messages are useful, once you learn what they mean. The following session contains examples of most of the common syntax errors, and the messages that result.

Forgetting the multiplication symbol is a common error.

```
> 2x + 3;
syntax error:
2x + 3;
 ^
```

The caret (^) indicates where Maple thinks the error is. Typing the wrong number of brackets is another syntax error.

```
> sin(exp(x);
syntax error:
sin(exp(x);
          ^
```

Using the wrong sort of quotes (see section 1.4.3) can cause difficulty:

```
> 'This is not a string, because the wrong quotes are used.';
syntax error:
'This is not a string, because the wrong quotes are used.';
          ^
```

Splitting an indivisible object across lines without using the continuation character (\ ) causes a syntax error.

```
> 1234
> 5678;
syntax error:
5678;
   ^
```

Instead, that should have been the following.

```
> 1234\
> 5678;
```

$$12345678$$

Some 'syntax errors' produce legal Maple code with a different meaning from that intended. Since Maple does not complain about this type of 'syntax error', they can be much harder to find. For example, incorrect capitalization will produce legal Maple code, as in the following example.

```
> SIN(x);
```

$$\mathrm{SIN}(x)$$

```
> diff(", x);
```

$$\frac{d}{dx} \mathrm{SIN}(x)$$

Unless the user has previously defined his or her own function SIN, he or she probably meant for Maple to differentiate $\sin x$ to get $\cos x$; the above result might convince the user that Maple was stupid. In fact, Maple's name for the sin function is sin, not SIN. But an error message would be inappropriate for the above session since the user might really want to talk about some other function called SIN.

1.1.5 Interrupting a Maple computation

Maple will sometimes 'go away' for quite a while to do its calculations. You may wish to interrupt the calculation, rather than wait for Maple to finish. This can usually be done, although since Maple makes complex use of memory it is not always possible to stop computation instantly. On DOS systems, pressing CTRL-Break will interrupt the calculation; on other systems it is CTRL-C, and on Unix it is the Unix interrupt character (for many users, *break*). For windowing systems, click on the interrupt button in the window. Execute the following command, and interrupt it before it finishes.

```
> int(1/(t^100-1), t);
```

Check your local documentation for the precise interrupt key if these do not work, but be aware that Maple will not always respond immediately to an interrupt.

1.2 Saving work

Save your work with the `save` command, which saves your variables and procedures, together with any library functions you have loaded in your session. In the following example session, several typical Maple commands are issued, followed at the end by two `save` statements that demonstrate the main ways this command is used. The sample Maple session itself should be reasonably intelligible even though you may not know the meaning of the commands used, so just skim it and pay particular attention only to the `save` statements.

```
> with(linalg):
> A := randmatrix(3, 3);
```

$$A := \begin{bmatrix} -85 & -55 & -37 \\ -35 & 97 & 50 \\ 79 & 56 & 49 \end{bmatrix}$$

```
> p := charpoly(A, x);
```

$$p := x^3 - 61\,x^2 - 9459\,x + 121529$$

```
> r := fsolve(p, x, complex);
```

$$r := -78.73703371, 12.09196353, 127.6450702$$

```
> readlib(realroot):
> root_bounds := realroot(p, 1/100000);
```

$$root\_bounds :=$$

$$\left[\left[\frac{1584917}{131072}, \frac{792459}{65536} \right], \left[\frac{8365347}{65536}, \frac{16730695}{131072} \right], \left[\frac{-10320221}{131072}, \frac{-2580055}{32768} \right] \right]$$

```
> evalf(");
```

$$[[12.09195709, 12.09196472], [127.6450653, 127.6450729],$$
$$[-78.73703766, -78.73703003]]$$

```
> save A, p, `matrix.mpl`;
> save `wholeworks.m`;
```

See the help entries for linalg, fsolve, realroot, etc., for more information on what those commands are capable of.

The command save A, p, `matrix.mpl` saves only the matrix A and the polynomial p in the human-and-Maple-readable file matrix.mpl. The command save `wholeworks.m` saves *everything*, including the library routines used by the linear algebra package that we loaded in, into the Maple-readable file wholeworks.m. The file extension tells Maple which format to use. The .m format is not human-readable but is more efficient for Maple to read (except that there are more variables and programs stored in it for this example). The left quotes (`) are different from right quotes (') and are necessary for the filenames so that the Maple *concatenation operator* period (.) doesn't turn the filenames into matrixmpl and wholeworksm, respectively. I consistently use the file extension .mpl to indicate a Maple input file. This convention is, of course, my own preference, and not a Maple requirement.

1.3 Use of files

The use of files for input is not mandatory in Maple. You may, if you so desire, continually retype commands again and again until you get them right. [It isn't reasonable to expect that the first attempt to solve a problem will be error-free.] It is remarkably more efficient (on a non-windowing system) to type commands *once* into a file, using your favourite editor, and then use the read command (or Unix I/O redirection) to read the file into Maple. Thereafter, make small modifications to the file to get the commands to do what you want. To make the read command work most effectively, the first line in your file should be

```
interface(echo=2);
```

The command interface(echo=2) tells Maple to echo the input together with the output: thus we get a sensible simulation of a Maple interactive session.

This command can be the last line in your .mapleinit file. This is the file that is automatically read in each time you start up Maple. Its name is system dependent: on Unix systems it is .mapleinit, on DOS systems its name is `maple.ini`, and it is different on other systems. Consult your local wizard or your system documentation for the exact name.

The reason this setting of the `echo` option is not the default is because the designers of Maple want to run their test files without seeing the input echoed. This setting conflicts with my personal style of Maple programming, and with that of other Maple programmers I know, and so I recommend that you change this setting in your `.mapleinit` file. I recognize, however, that you may wish to use fewer Maple 'scripts' (i.e., files of Maple commands) and more 'procedures', as the developers of Maple do, in which case the default setting of `echo` may suit your style better than the setting I recommend.

Bill Bauldry informs me that many versions of Maple require the initialization file to end with a blank line and carriage return. In any event this ending of the file can do no harm.

Exercise

1. On your system, write your initialization file `maple.ini` (or `.mapleinit` or whatever the appropriate name is). Include `interface(echo=2)` as your last command. Put a command in to print out a message showing that initialization has happened.

1.3.1 Worksheets

The concern of the previous section is less important now than it used to be, because the advent of windowing systems is changing the most natural way of using Maple from this 'batch-oriented' method to a 'worksheet-oriented' method. The one problem with worksheets that I have noticed is that if you do not stick to a strict, top-down order of execution of commands, you may not be able to reuse the worksheet. Otherwise, they seem much superior, both in the aesthetic quality of the output and in the ease of their use.

One feature I find especially handy is the ability (new to Release 3) to save in LaTeX format. This allows you to convert your worksheet to a LaTeX input file for text processing. Much of this book was prepared (or at least polished) using this feature.

1.3.2 Writing to a file

Save your printed output with the menu command `Print Session Log` if this is available on your system, save your worksheets with the `Save` menu item if this is available on your system, or save your output by redirecting output (in the normal way on a Unix system or) using the `writeto` or `appendto` commands. These last two commands must be issued before you do your work.

1.4 Some things to watch out for

Your previous experience with computers may have taught you to expect certain patterns—some of these can lead you astray in Maple. In particular, variables can have a much wider variety of values in Maple than they can in FORTRAN.

1.4.1 Assigning values to variables

First, if an identifier (say x) has not been assigned a value, then in Maple it stands for itself—that is, it is a *symbol*. This is different from FOR-TRAN, where reference to an undefined variable may get you a NaN (Not a Number), zero, an error message, or garbage, depending on the compiler.

In Maple, symbols (variables) are assigned values with := (pronounced 'becomes') as in Pascal and *not* with = as in C or FORTRAN. It is common to confuse the two.

```
> p := x^2 + 3*x + 7;
```

$$p := x^2 + 3\,x + 7$$

```
> x;
```

$$x$$

Note that x has no value (it represents itself) and that p has been assigned a value.

```
> p;
```

$$x^2 + 3\,x + 7$$

```
> q = p;
```

$$q = x^2 + 3\,x + 7$$

That looks like assignment, but it is an *equation*.

```
> q;
```

$$q$$

Nothing happened to q as a result of the previous statement. You can assign equations to variables:

```
> r :=  q = p;
```

$$r := q = x^2 + 3x + 7$$

The variable r has the value $q = x^2 + 3x + 7$. This kind of value for a variable is not possible in a purely numerical language such as FORTRAN.

1.4.2   Removing values from variables

A candidate for the most common cause of bugs is *regarding a variable that has a value as a symbol*. For example, suppose that in the first part of your session, you set x := 3, and much later you define p := x^2. Then suppose you try to integrate p with respect to x, having forgotten that x has a value; you wind up asking Maple to integrate with respect to 3, which doesn't make sense. We will see an example of this shortly. You can 'unassign' variables by a statement of the form x := 'x'. See also the command unassign.

```
> p := x^2 + sin(x);
```

$$p := x^2 + \sin(x)$$

```
> x := 3;
```

$$x := 3$$

Now suppose we do other things for a while, and forget that $x = 3$. Now we want to integrate p.

```
> int(p, x);
Error, (in int) wrong number (or type) of arguments
```

This error message is not terribly illuminating to the uninitiated.

```
> x := 'x';
```

$$x := x$$

```
> int(p, x) + C;
```

$$\frac{1}{3}x^3 - \cos(x) + C$$

Order is important up there: let's try assigning x, then p.

```
> x := 3;
```

$$x := 3$$

```
> p := x^2 + sin(x);
```

$$p := 9 + \sin(3)$$

```
> int(p, x);
Error, (in int) wrong number (or type) of arguments
```

Again, we must interpret that error message (it means the same as it did last time—we are trying to get Maple to integrate with respect to 3).

```
> x := 'x';
```

$$x := x$$

```
> int(p, x);
```

$$(9 + \sin(3))x$$

Now what happened?

In the last part, we assigned a value to p after we had assigned a value to x. Full evaluation of the right-hand side takes place, so in the second part, p was assigned the value $9 + \sin(3)$, with no more links or references to x. Hence the computed answer.

1.4.2.1   Recursive definition of name

One other common error is the following.

```
> x := x + 1;
Warning, recursive definition of name
```

$$x := x + 1$$

Maple warns us about that assignment, and we will see why. In FORTRAN such an assignment makes sense, because you can only refer to variables that already have values—so this would be an increment or update to the value that x has. In Maple, though, x will usually not have a value at all—so now x is defined in terms of itself. If we ask Maple to give us sin x, what happens?

```
> sin(x);
Error, STACK OVERFLOW
```

Maple tried to find sin x; it needed to know what x was—and $x = 1 + x = 1 + (1 + x) = 1 + (1 + (1 + x)) = \cdots$, which never terminates. So Maple runs out of stack space trying to evaluate x.

Watch out for defining p in terms of q and q in terms of p, also.

Exercise

1. Set x := 5; and then try x := x + 1; and explain what you see. This feature allows updating of indices in loops, for example.

1.4.3   Left quote vs. right quote vs. ditto

The distinction between the different types of quotes is important, and many more examples of their use will be given in this book. To get help from Maple on the different quotes, type ?quotes.

Left quotes are string delimiters.

```
> `This is a complicated string with spaces`:
```

Right quotes prevent evaluation.

```
> 'sin(Pi)';
```

$$\sin(\pi)$$

```
> ";
```

$$0$$

The double quotes (") refer to the previous result, while 'double' double quotes ("") refer to the result before that, and 'triple' double quotes (""") refer to the result before that. If you wish to refer to still earlier results, you must have previously issued the history command, or named the results in an assignment statement. Otherwise, you are out of luck. It is often possible to re-execute input commands using the 'history' mechanism of your operating system—for example, on a DOS machine simply hitting the up-arrow will recover the previous command, and hitting it again will get the one before that, and so on. Re-executing commands is usually less efficient than saving the results the first time, but also may take less time than the first computation did because many Maple commands *remember* their previous workings.

See section 3.4 for more discussion on the double-quote variables.

Strings in Maple are either *simple strings*, which are sequences of one or more letters, digits, and underscores beginning with a letter, or *quoted strings*, which are any sequence of characters (including newlines) enclosed in left quotes ('). To enclose a left quote in a string you must double it up.

```
> This_is_a_simple_string_and_no_quotes_are_necessary;
              This_is_a_simple_string_and_no_quotes_are_necessary
```

```
> 'Since this string contains spaces and commas, quotes are necessary.';
        Since this string contains spaces and commas, quotes are necessary.
```

If a quoted string contains a newline character, Maple warns us.

```
> 'This string
Warning, string contains newline character. Close strings with ' quote.
```

```
> contains a newline.  Notice that Maple warns us about that.';
          This string
          contains a newline.  Notice that Maple warns us about that.
```

Here is an example of what could happen if no warnings were issued.

```
> 'That is because of the following. [I deleted the warnings
> so you would see what would happen if there were none.]
> int(sin(x), x);
> Wait---why didn't my command work?  Because we are still
> in the string!  We can have left quotes '' in this string,
> but we have to double them up.';
          That is because of the following. [I deleted the warnings
          so you would see what would happen if there were none.]
          int(sin(x), x);
          Wait---why didn't my command work?  Because we are still
          in the string!  We can have left quotes ' in this string,
          but we have to double them up.
```

Complicated strings can be formed from other strings by using the period (.) operator, which is the Maple concatenation operator. A common example of this is the creation of several related names a1, a2, etc., although this may be more usefully done with indexed names. The following introduces the notion of a *range* in Maple. A range is simply two integers connected with two periods, as follows, and it means all the integers from the lower to the upper value, inclusive.

```
> a.(1..5);
```

$$a1, a2, a3, a4, a5$$

```
> p := b0;
```

$$p := b0$$

```
> for i to 5 do p := p +b.i*x^i; od:
> p;
```

$$b0 + b1\,x + b2\,x^2 + b3\,x^3 + b4\,x^4 + b5\,x^5$$

Any Maple string can be used as a variable name. This can sometimes look a little strange, and so I recommend that only simple strings be used as variable names, following common practice. Here is a 'strange' example, to show what I mean.

```
> '2 + 2' := 5;
```

$$2 + 2 := 5$$

An interesting exception is the following. The variable name below contains a special character which is interpreted by LaTeX as a superscript—thus one can create special effects in the LaTeX output. I have not thought of a use for this trick, but it might come in handy. I had to strip a spurious italics font command (generated automatically by the LaTeX output facility) out of the variable name, though, so the power would print legibly.

```
> '(x+y)^4' := (x+y)^2;
```

$$(x + y)^4 := (x + y)^2$$

1.4.4   Precedence of operators

The Maple arithmetic operators are + for addition, * for multiplication, – for subtraction and negation, / for division, mod for modular arithmetic, and ^ for exponentiation. Maple will also allow use of ** for exponentiation, if you prefer it.

The precedence of these operations is the natural one, where exponentiation takes place first, followed by multiplication and division, and finally by addition and subtraction. Arithmetic expressions are read left-to-right, subject to parentheses and the above precedence rules, as is standard in most computer languages. Thus 1 + 2*3^3 produces 55 and not 729.

Also, the operator &* was taken from the list of 'inert' operators (which are programmable binary operators) to represent the noncommutative matrix multiplication operator. *Do not try to multiply matrices with* *.

In Maple V Release 3 and earlier, &* has the same precedence as *, and the other inert operators &+, &-, and &/ all have higher precedence. This has non-intuitive results when these operators are mixed in a single expression, so parentheses must be used: a &+ b &* c is interpreted by Maple as (a &+ b) &* c. If you mean a &+ (b &* c) you must say so.

1.4.5 Protected and reserved names

You may not use Maple reserved words as variable names: for example, if, fi, do, od, next, and other control structure names. Maple will immediately object to the use of reserved words as variable names, with syntax errors.

Some other names are not 'reserved' as part of the Maple programming language, but are 'protected'. These are usually names of crucial pieces of library or kernel code, such as op, Pi, and D. You *can* undo this protection by issuing the command unprotect(Pi);, but this is not recommended. See ?unprotect for more details.

Finally, some names (such as I, Maple's name for the square root of -1) are implemented via alias and cannot be assigned to until they are aliased back to themselves (see section 1.8.0.1).

This discussion skirts around a larger issue. I consider the protect command a step forward from the previous state, but I think that the Maple user should be allowed to use whatever single-letter names she or he wishes, for whatever purpose: currently, the single-letter names C, D, I, W, Pi, gamma, GAMMA, Zeta, and many others, are forbidden to the user, to varying degrees. It is usually possible to use C with impunity, and impossible to use D without 'unprotecting' it and risking catastrophic failure of Maple. There is ongoing debate within the Maple group on how best to resolve this issue.

1.5 Documenting your work

Choose your variable and procedure names carefully. They should be neither too short (= too cryptic) nor too long (= too unwieldy). They should instead be *appropriate*. Calling a list of constants Fred after your friend might be amusing but calling it List_of_Constants is much more understandable. Of course, that is hard to type, so listcon might be better still (though with better editors and macros there is less excuse for such laziness today—List_of_Constants is clearly to be preferred for infrequent use). Use longer names for global variables to minimize conflicts with other pieces of code. Use local and environment variables where possible. [See section 3.4 for definitions of these.] Use `macro` and `alias` to allow multiple names for the same object.

Pick a consistent programming indentation style for your loops. Reading your own programs two months later, you will be grateful. In Maple, there may be many ways of expressing the same action, but pick one and use it consistently.

Comment your work. The comment character in Maple is `#`. Anything after this character on a line is unseen by Maple. Useful comments tell how to use a particular piece of code, what is expected as input, and what the output will look like. Next most useful is a maintenance history. This lets you compare two copies of code and use the most recent. Many hard disks come to resemble the shells of ancient marine creatures, with old, fossilized bits of code and data hanging about. A maintenance history helps deal with this. Finally, comments giving references to the papers or books whence the algorithms came, or discussing how the algorithms work, are also worthwhile.

For a procedure, write a help file for it also. This doesn't have to be large and can simply be the usage information mentioned earlier, together with a reference and an example.

For a package, write a help file for the package and a help file for each user-level routine.

For a large script file that uses many files meant to be read in and executed in a particular way, write a README file describing the overall structure and purpose.

Of course all this increases your workload, but greatly improves your total throughput and productivity, as much less time is wasted re-inventing your own wheels.

Consider the following example program, which swaps the values of two variables.

```
#
# swap:  Procedure to swap variables: RMC April 1994

# Basic idea:  quote variables in the *call* so that
#              the procedure knows where to put values.
#
# Arose in discussion with George Corliss.

# MAINTENANCE HISTORY:
#
#  First version utterly wrong and thrown away.
#  This version April 15, 1994.

# CALLING SEQUENCE:  swap('x', 'y');

# INPUT: a,b  : names of variables containing values.

# OUTPUT: the NULL value

# SIDE EFFECTS:  exchanges the values of the actual arguments
#               corresponding to 'a' and 'b'.
```

```
# GLOBAL VARIABLES:  none at all
# KNOWN BUGS/WEAKNESSES:  If 'x' has the value 'y' then
#                         a recursive definition of name can occur
#                         after the call swap('x', 'y'); caveat emptor.
#
#       Patient:  It hurts when I do this. <raises right arm and squints>
#       Doctor :  Well, don't do that.  <raises right arm and squints>
# Miscellaneous remarks:  Use of 'eval' is necessary to prevent
#                         recursive definition of name---one-level
#                         evaluation of 'a' and 'b' would result
#                         in x being assigned y and y being assigned
#                         x, which will cause a memory fault on
#                         attempted evaluation of x or y.
swap := proc(a:name, b:name) local temp;

  if not(assigned(a) and assigned(b)) then
    ERROR('Variables must have values to swap');
  fi;

  temp := eval(a);
  a := eval(b);
  b := temp;

NULL

end:
```

Observe the following points:

1. Names are not too long (`swap`, `temp`) but are intelligible. Use of single-letter names is appropriate for mathematical variables or symbols.

2. The procedure checks its input, both in the formal parameter list (see Chapter 3) and afterwards to see if the names have values.

3. The comments take roughly three times the space of the program. Now, that's partly because I am a wordy person, and it is easier for me to write a lot than to write a little. It is possible to boil the comments down to a bare minimum, and indeed this can be a useful exercise. But wordiness doesn't hurt, here, and the information given can be helpful to explain the code to others, or to yourself after enough time has passed.

4. The comments include some historical information. Typically this would include a reference for the algorithm (but that would be silly in this case).

5. White space (blank lines and spaces) and consistent indentation are used to make the code easier to read.

6. The calling sequence, the expected input, the expected output, and any side effects are the most important items to document.

1.6   The three levels of Maple 'black boxes'

Maple provides many 'black boxes' for use in mathematical applications. Some of these are built-in and available immediately, such as `series` or `int`. We will refer to such routines as 'main' routines. The first of the above examples, `series`, is written in C and is part of the Maple 'kernel'. The second of these is written in the Maple programming language and is part of the built-in library. It is not really present in memory just after Maple has been started—the initial definition of `int` is `'readlib('int')'`, which causes the main part of the library routine for `int` (which is actually a very small part of the overall program) to be read in the first time `int` is invoked. No reading occurs on any subsequent invocation, because `readlib` has option `remember` which we will discuss later. The `readlib` command greatly increases the memory efficiency of Maple, because unused facilities are not loaded. But in effect, `int` is available whenever it is needed by the user, and so it is classified here as a main routine.

There are library routines that are not 'readlib-defined' in this way, and the user has to explicitly load them in via a call to `readlib`. For example, the routine `sturm`, which computes the number of real roots of a polynomial in an interval, must be loaded by such an explicit call. I will call such routines 'readlib' routines.

```
> readlib(sturm):
> ?sturm          #Do this to see the help page for sturm
> sturmseq(x^3-2*x+1, x);
```

$$\left[x^3 - 2\,x + 1,\, x^2 - \frac{2}{3},\, x - \frac{3}{4},\, 1 \right]$$

```
> sturm(", x, 0, 2);
```

$$2$$

Hence there are two real roots x in the interval $0 < x < 2$.

The third type of 'black box' available in Maple comes in a 'package', such as the `linalg` package. A package is a collection of more or less related routines that is loaded into Maple by the user with the `with` command. For example, in the following session we load and use a routine from the student package.

```
> ?packages
> with(student);
```

[*D, Doubleint, Int, Limit, Lineint, Sum, Tripleint, changevar, combine, completesquare, distance, equate, extrema, integrand, intercept, intparts, isolate, leftbox, leftsum, makeproc, maximize, middlebox, middlesum, midpoint, minimize, powsubs, rightbox, rightsum, showtangent, simpson, slope, trapezoid, value*]

```
> ?changevar      # Do this to see the help page for changevar
> Int( sin(x^2)*x, x= -1..1);
```

$$\int_{-1}^{1} \sin(x^2) \, x \, dx$$

```
> changevar(u=x^2, ", u);
```

$$\int_{1}^{1} -\frac{1}{2} \sin(u) \, du$$

```
> value(");
```

$$0$$

The computed value of that integral in the last example is correct—the original integrand was odd and integrated over a symmetric interval—but changevar paid no attention to the legality of the change of variable ($u = x^2$ has zero derivative in the range of integration and so the theorem guaranteeing correctness of the change of variable does not apply). Maple got this one right accidentally.[1] This brings up an extremely important point:

1.7 No non-trivial software package is bug-free

Maple has bugs. It has *always* had bugs. Though it is an evolving system and bug-fixing is a major activity of the Maple group, it always *will* have bugs. The same statements hold for *any* computer language, and indeed for any program I have ever used (even MATLAB). The designers of Maple take the pragmatic view that it is better to have an actual program available for people to use, bugs and all, than to have a perfect program on the drawing board, or 'ready any day now.'

Every other computer algebra system also has bugs, often different ones, but remarkably many of these bugs are seen throughout all computer algebra systems, as a result of common design shortcomings. Probably the most useful advice I can give for dealing with this is *be paranoid*. Check your results in at least two ways (the more the better). Don't just do the calculation again in the same way. Instead, do a simple case by hand, do selected cases numerically, substitute your solution back into the defining equation and look at the residual, do the same for the initial and boundary

---

1. To be fair, 'changevar' is intended to let the user do what she or he wants.

conditions, plot the results, and compare with physical experiment and the results of other people's calculations.

This will help you correct bugs in Maple, bugs in your own code, and bugs in your problem setup. It may even help you correct bugs in your thoughts.

When you find a bug in Maple, report it to `support@maplesoft.on.ca` by e-mail or phone it in to Waterloo Maple Software. If you can, isolate the bug in as small a code fragment as possible, and clearly explain what you think should be happening. The best bug reports include their own suggested fixes, and these get the highest priority.

Remark. If Maple is so buggy, why do I recommend that people use it? And why do I use it so much myself? Nearly all my research work uses Maple, sometimes only for trivial purposes but sometimes to make the central investigations of the paper. My point is that skepticism is healthy, not that Maple is unusable. Obviously, the opposite is true—Maple is very usable indeed, and sometimes indispensable. But people who uncritically believe what computers tell them deserve what they get: Garbage In, Gospel Out is a poor motto for our age.

Exercises

1. Find out how to print a Maple library procedure so you can look at it. Look at the code for `series`, `int`, and `latex`. Comment on your investigations. [Hint: `?interface` and `?print`.]

2. Give a definition of a 'bug in the program.'

3. The result of a successful `readlib` command is the body of the procedure you wanted to read in. What do you predict will happen in the session below? Explain.

   ```
   > p := 1 + 2*x + x^2:
   > readlib(realroot):
   > "("", 1/1000);
   ```

1.8 Evaluation rules

Maple's evaluation rules are designed to do what the user will expect in most situations. However, that is impossible—different people have different expectations in the same situations. The `eval` function and right quotes (') are used to customize evaluation.

1. Global variables are evaluated fully, in most cases. For example,

   ```
   > x :=y;
   ```

$$x := y$$

```
> y := z;
```

$$y := z$$

```
> z := 3;
```

$$z := 3$$

```
> x;
```

$$3$$

```
> y;
```

$$3$$

Quotes prevent evaluation—a single evaluation merely removes the quotes and does not go 'all the way down.'

```
> 'x';
```

$$x$$

That is why the following construct allows you to 'unassign' a name, or clear a variable.

```
> x := 'x';
```

$$x := x$$

```
> x;
```

$$x$$

```
> y;
```

$$3$$

2. Local variables are evaluated *one* level in procedures. This is usually what is wanted.

```
> f := proc(q) global x;
>    x := 5;    # Assignment to a global variable
>    q          # Return the 1-level evaluation of q
> end:
> f(x^3 + x + 1);
```

$$x^3 + x + 1$$

```
> g := proc(p) global x;
>    x := 5;
>    eval(p);   # Fully evaluate p this time
> end:
> g(x^3 + x + 1);
```

Note the explicit declaration of the variable x as global. This declaration is necessary in Maple V Release 3, and impossible in previous releases. See Chapter 3 for a further discussion of global, local, and environment variables.

3. Arrays and tables are evaluated to the *last name*. For example,

```
> T := table():
> T[jumper] := climber;
```

$$T_{jumper} := climber$$

```
> T[spelunker] := abseiler;
```

$$T_{spelunker} := abseiler$$

```
> T[1] := 5;
```

$$T_1 := 5$$

```
> T;
```

$$T$$

```
> eval(T);
```

$$\text{table}([$$
$$jumper = climber$$
$$1 = 5$$
$$spelunker = abseiler$$
$$])$$

```
> with(linalg):
> A := matrix(3, 3, (i,j) -> 1/(x + i + j) );
```

$$A := \begin{bmatrix} \dfrac{1}{x+2} & \dfrac{1}{x+3} & \dfrac{1}{x+4} \\ \dfrac{1}{x+3} & \dfrac{1}{x+4} & \dfrac{1}{x+5} \\ \dfrac{1}{x+4} & \dfrac{1}{x+5} & \dfrac{1}{x+6} \end{bmatrix}$$

```
> A;
```

$$A$$

```
> eval(A);
```

$$\begin{bmatrix} \dfrac{1}{x+2} & \dfrac{1}{x+3} & \dfrac{1}{x+4} \\[2mm] \dfrac{1}{x+3} & \dfrac{1}{x+4} & \dfrac{1}{x+5} \\[2mm] \dfrac{1}{x+4} & \dfrac{1}{x+5} & \dfrac{1}{x+6} \end{bmatrix}$$

print(A); will also display the contents of A, though unlike eval it does not make the visible results accessible to subsequent Maple statements; rather, print just prints to the screen.

4. Objects are *not* evaluated by the subs or subsop commands. For example,

```
> f := sin(x);
```

$$f := \sin(x)$$

```
> subs(x=0, f);
```

$$\sin(0)$$

5. Evaluation can be prevented with right quotes (') and forced with eval as seen previously.

```
> for i to 5 do i^2; od;
```

$$1$$

$$4$$

$$9$$

$$16$$

$$25$$

```
> i;
```

$$6$$

```
> sum(i^3, i=1..n);
Error, (in sum) summation variable previously assigned,
                second argument evaluates to, 6 = 1 .. n
```

```
> sum('i^3', 'i'=1..n);
```

$$\frac{1}{4}(n+1)^4 - \frac{1}{2}(n+1)^3 + \frac{1}{4}(n+1)^2$$

The above examples show that the eval command can be used to force the desired level of evaluation in exceptional circumstances. (See also (10), pp. 75–76.)

For special domains, such as matrices, hardware floating point, Maple floating point, complex numbers, or others, special evaluation routines are used, as follows.

The routine evalm is used for matrices.

```
> with(linalg):
> eye := n -> array(1..n, 1..n, identity):
```

Alternatively, &*() is the identity matrix for any n.

```
> A := matrix(2, 2, (i,j)->1/(i+j+x)):
> B := evalm(A + 6*eye(2));
```

$$B := \begin{bmatrix} \dfrac{1}{2+x} + 6 & \dfrac{1}{3+x} \\ \dfrac{1}{3+x} & \dfrac{1}{4+x} + 6 \end{bmatrix}$$

Compute the inverse of A.

```
> evalm(A^(-1));
```

$$\begin{bmatrix} (2+x)(3+x)^2 & -(3+x)(2+x)(4+x) \\ -(3+x)(2+x)(4+x) & (4+x)(3+x)^2 \end{bmatrix}$$

The matrices A and B commute.

```
> evalm(A &* B - B &* A);
```

$$\begin{bmatrix} 0 & 0 \\ 0 & 0 \end{bmatrix}$$

```
> p := collect(charpoly(A, lambda), lambda, factor);
```

$$p := \lambda^2 - 2\,\frac{(3+x)\lambda}{(2+x)(4+x)} + \frac{1}{(2+x)(4+x)(3+x)^2}$$

Verify that A satisfies its characteristic polynomial.

```
> evalm(subs(lambda=A, p)):
> map(normal, ");
```

$$\begin{bmatrix} 0 & 0 \\ 0 & 0 \end{bmatrix}$$

The routine `evalhf` is used for hardware floating-point evaluation (for speed).

```
> evalhf(sin(1));
```

$$.8414709848078965$$

The routine `evalf` is used for Maple arbitrary-precision floating point (for more accuracy than `evalhf`).

```
> Digits := 20;
```

$$Digits := 20$$

```
> evalf(sin(1));
```

$$.84147098480789650665$$

Alternatively, you can pass the number of requested digits to `evalf` in the argument list.

```
> evalf(sin(1), 40);
```

$$.8414709848078965066525023216302989996226$$

Further, floats are *contagious* to some extent, so `sin(1.)` automatically evaluates to $0.8414\ldots$.

The routine `evalc` is used for complex numbers.

```
> x := a + b*I;
```

$$x := a + bI$$

```
> y := c + d*I;
```

$$y := c + dI$$

```
> x*y;
```

$$(a + bI)(c + dI)$$

```
> evalc(");
```

$$ac - bd + (ad + bc)I$$

```
> evalc(Re("));
```

$$ac - bd$$

```
> evalc(Im(""));
```

$$ad + bc$$

```
> evalc(Re(1/x));
```

$$\frac{a}{a^2 + b^2}$$

Note that `evalc` assumes implicitly that names are *real-valued*, while most of the rest of Maple does not.

1.8.0.1 Why is 'I' Maple's name for the square root of -1?

Maple is also a programming language, and since the designers of Maple are primarily programmers, their concerns are those of programmers. Since lower-case i is used extensively as an index in `for` loops, as is j, the Maple designers could not bear to use either for the imaginary unit. They had no such need for upper-case I, and some also have an aversion to using two-character codes such as `_i`. What is needed is a mathematically typeset math-italic i—but that isn't available in ASCII.

The preempting of I for the imaginary unit causes some difficulty if you wish to use I as a name for an integral, or as an identity matrix or operator, etc. However, you *can* do it, since I is now predefined in Maple using `alias` instead of assignment. If you want to use the variable I for your own purposes, you issue the command `alias(I=I)` first, which frees it for your own use. In the following, suppose that A is a predefined 2 by 2 matrix. Then we can use I as the identity matrix as follows.

```
> alias(I=I):
> I := array(1..2, 1..2, identity):
> evalm(A + lambda*I):
```

1.8.1 Inert functions

Sometimes you don't want Maple to evaluate an expression at all, or at least until you tell Maple to do so later. A useful mechanism for doing this is provided by the concept of an 'inert function.' Typically, the name of an inert function in Maple is the same as the active function it represents, except that the initial letter is capitalized: `Int`, `Sum`, `Diff`, `Svd`, and `Eigenvals` are examples. This nomenclature rule is not universal, and exceptions include `RootOf` and `DESol`.

A typical use of an inert function call is to prevent symbolic evaluation of the problem and allow later numerical procedures to be invoked. This can save time spent symbolically processing the input in the case where such processing fails. Consider the following example, where we first try to integrate a function that has no closed-form antiderivative. We time the results using `time()`. The following is the standard syntax for the use of this routine, but apparently we can more simply use `time(<expression>);` to compute the time to evaluate the inner expression. This feature is undocumented and I have not used it yet, and so I have left my examples as they are. I plan to test the new syntax shortly, as it is clearly simpler. Now back to the example.

```
> st := time();   evalf(int( exp(-t)*arcsin(t), t=0..1));
> time_taken := time() - st;
```

$$st := 1.000$$

$$.2952205678$$

$$time\_taken := 52.000$$

That took 52 seconds to get a numerical answer. Some portion of that time was spent in symbolic processing, deciding that $\int \exp(-t) \sin^{-1}(t)\, dt$ cannot be done. If we restart the session to remove all possible effects of `option remember` and redo the calculation with an inert `Int`, we see a noticeable improvement.

```
> restart;
> st := time();   evalf(Int( exp(-t)*arcsin(t), t=0..1));
> time_taken := time() - st;
```

$$st := 54.000$$

$$.2952205678$$

$$time\_taken := 22.000$$

So, in the previous try, more than half the time was spent on symbolic processing. Of course, if it had succeeded, then usually the resulting numerical evaluation would be more efficient.

Let us consider a case where symbolic processing *must* fail—computation of the eigenvalues of a general five by five matrix.

```
> A := linalg[randmatrix](5, 5);
```

$$A := \begin{bmatrix} -85 & -55 & -37 & -35 & 97 \\ 50 & 79 & 56 & 49 & 63 \\ 57 & -59 & 45 & -8 & -93 \\ 92 & 43 & -62 & 77 & 66 \\ 54 & -5 & 99 & -61 & -50 \end{bmatrix}$$

We do not bother to try `eigenvals` on this, because it will just return a `RootOf` containing the characteristic polynomial. It is well known that finding the roots of polynomials is much harder than finding the eigenvalues of the matrix directly (44).

```
> evalf(Eigenvals(A));
```

$$[-103.5826423, \quad 75.17618049 + 100.5891723I, \quad 9.615141005 + 78.23304474I,$$
$$75.17618049 - 100.5891723I, \quad 9.615141005 - 78.23304474I]$$

I rearranged the output a little bit to make it more readable. Similarly for the singular values below.

```
> evalf(Svd(A));
```

$$[215.3159815, 192.8527895, 110.7852504, 66.73636029, 33.05625793]$$

Similar considerations hold for evaluating sums as held for evaluating integrals. If we call `sum`, we are asking for closed-form anti-differences (see section 2.4), which can take a long time to fail. If what we want is a numerical sum, it is best simply to do as follows.

```
> Sum(1/k^2, k=1..infinity);
```

$$\sum_{k=1}^{\infty} \frac{1}{k^2}$$

```
> evalf(");
```

$$1.644934067$$

Of course, Maple knows that the sum is $\pi^2/6$ and symbolic processing would succeed here. Consider now the following counterintuitive result.

```
> Sum(1/sqrt(k), k=1..infinity);
```

$$\sum_{k=1}^{\infty} \frac{1}{\sqrt{k}}$$

```
> value(");
```

$$\infty$$

The command `value` converts an inert function to its active form, forcing symbolic processing. Here Maple correctly determines that the series diverges to ∞. What does `evalf(Sum(...))` do?

```
> evalf("");
```

$$-1.460354509$$

It returns a negative answer! This is because the sequence acceleration method used (Levin's u-transform) will sometimes give a numerical value to a formally divergent series. This is often useful, because formally divergent sums may (with appropriate convergence acceleration techniques) sum to physically interesting values. Here, adding positive terms to get a negative value does not seem to make sense—but this can be justified via analytic continuation (of the Riemann ζ-function—evaluate $\zeta(1/2)$ in Maple to see how close the above answer is). It is clear that you must use `evalf(Sum)` with caution.

Here is another example of inert functions, showing conversion to the active form by using `value`.

```
> Int( ln(x)/(1-x^2), x);
```

$$\int \frac{\ln(x)}{1-x^2}\, dx$$

```
> value(");
```

$$\frac{1}{2}\,\mathrm{dilog}(x) + \frac{1}{2}\,\mathrm{dilog}(x+1) + \frac{1}{2}\ln(x)\ln(x+1)$$

Finally, here is an example showing inert functions used to evaluate functions over a finite field.

```
> p := x^5 + x^3 + x;
```

$$p := x^5 + x^3 + x$$

```
> Factor(p) mod 2;
```

$$x(x^2 + x + 1)^2$$

```
> q := diff(p, x) mod 2;
```

$$q := x^4 + x^2 + 1$$

```
> Gcd(p, q);
```

$$\mathrm{Gcd}(x^5 + x^3 + x, x^4 + x^2 + 1)$$

```
> " mod 2;
```

$$x^4 + x^2 + 1$$

Exercises

1. If $z = x + iy$, where x and y are real numbers, use Maple to find $\Re[(z+1)/(2+z^2)]$. (Note: $\Re(z)$ is the real part of z.)

2. If A is the matrix below, find $I + A + A^2 + A^3$ using evalm. Note that it is obvious in this context that I is meant to be the identity matrix, not the square root of -1.

$$A = \begin{bmatrix} 1 & 1 & a \\ 1 & a & 0 \\ a & 0 & 1 \end{bmatrix}$$

3. Define a variable f to be the expression $x+\sin(2x)$. Use subs to evaluate this expression if $x = 0$, 1, and $-\pi$.

4. Compare the execution times of evalf(int(1/(1+t^12), t=0..1)) and the inert form. Explain.

5. Compare the execution times of evalf(sum(exp(-k), k=1..infinity)) with its inert form. Examine convert([seq(exp(-1.0*k), k=1..40)], '+') and the for-loop, also.

   ```
   > s:= 0: for i to 40 do s := s + exp(-1.0*k); od:   s;
   ```

   Be careful to use a colon (:) and not a semicolon (;) at the end of od:, because otherwise all the intermediate results will be printed, which will disrupt the timings.

6. Compute the singular values of a random 10 by 10 matrix with evalf(Svd(A)). Compare the execution speed with MATLAB, if you have access to it. MATLAB is designed to be very fast at this type of computation.

1.9 The assume facility

> 'I can be of little avail to your lordship if you give me
> unsufficient premises to reason from. But worse than tell it not to
> me, I fear you tell it not truly to yourself.'
> —E. R. Eddison, *A Fish Dinner in Memison*, p. 184.

Maple is a computer *algebra* language. It is now apparent that many people try to use Maple (or other computer algebra languages) for *analysis*. This has caused some difficulty in the past since there are differences between what is considered correct if you are doing algebra and what is considered correct if you are doing analysis. For example, consider the solution of $(k^4 + k^2 + 1)x = k^4 + k^2 + 1$. If k is real, then we can say unequivocally that $x = 1$, but if k might be one of the complex roots of

$k^4 + k^2 + 1$, we must add a proviso to the result $x = 1$ for complete correctness, namely that $x = 1$ provided $k^4 + k^2 + 1 \neq 0$. This is an example of a correct analytical result. However, the *algebraic* approach to this problem would be to consider k as an *indeterminate*. Thus k has *no* value, and so it is perfectly legal to divide by this polynomial in k, so long as it isn't the zero polynomial (which it isn't, since not all the coefficients are zero). In that case, $x = 1$ with no provisos at all. Computer algebra systems *often take this point of view*, and it is only recently that the analytical viewpoint has been considered in many cases.

A related problem is the evaluation of integrals and sums that depend on parameters. It is possible that for some values of the parameters, the integral is finite, and for others, that the integral does not converge. For example, consider

$$\int_0^\infty \exp(-st)\,dt\,.$$

In a naive attempt to solve this with algebra, we find the antiderivative for $\exp(-st)$, which is $-\exp(-st)/s$, and plug in the limits of the integration: $-1/s$ at the bottom and $\cdots$, well, something at the upper limit: $-\exp(-s\infty)$. It is easy to overlook the fact that $\Re(s)$ might be nonpositive and indeed there has not been, until recently, any way of telling Maple about this analytical, or geometric, information.

We need some way of telling Maple about our assumptions on s—in this case that (for example) $s > 0$. Once Maple knows about the properties of the variables, it can proceed *correctly*. Prior to Maple V Release 2, Maple would simply give you the answer $1/s$ for this integral, which would be incorrect if $s \leq 0$. Since Maple V Release 2, Maple makes some attempt to inform you that it needs to know more about s.

```
> int(exp(-s*t), t=0..infinity);
```

$$\lim_{t \to \infty-} -\frac{e^{-st}}{s} + \frac{1}{s}$$

and the user is supposed to infer from this unevaluated limit that Maple can do *something* with the integral but it needs to know more about s before it can evaluate the limit.

> 'Anthea laughed. "Timourous scrupulosities!
> 'Twere meant, if not said."'
> —E. R. Eddison, *A Fish Dinner in Memison*, p. 163.

This type of behaviour is not universal: for example, consider

```
> int( exp(-t)*t^(x-1), t=0..infinity);
```

$$\int_0^\infty e^{-t} t^{x-1} dt$$

which simply returns unevaluated. If you tell Maple that $x > 0$ before
issuing the above command, then Maple will simplify the above integral
to $\Gamma(x)$, the gamma function. Looked at one way, the above integral does
not converge if $x \le 0$, and so is not equal to $\Gamma(x)$ in that case; strict rigour
prevents Maple from giving an answer without the assumption that $x > 0$.
However, Maple is in an awkward position above, unable to tell the user
that it can do more with the proper assumptions. The user may well not
know that something further can be done with that integral. The idea of
provisos (16) is meant to deal with this issue, and this is under discussion
at the moment.

These problems are representative of the problems the `assume` facility
is meant to address. The following session illustrates some of the simple
things you can do with `assume`.

```
> ?assume
> assume(s>0);
> Int(exp(-s*t), t=0..infinity);
```

$$\int_0^\infty e^{-\tilde{s} t} dt$$

```
> value(");
```

$$\frac{1}{\tilde{s}}$$

The tilde (or twiddle) ˜ after the s here is Maple's way of reminding you
that you have made assumptions about s.

```
> int(exp(-a*t), t=0..infinity);
```

$$\lim_{t \to \infty-} -\frac{e^{-at}}{a} + \frac{1}{a}$$

The example below shows that Maple can figure out that $s + 1 > 0$ if it
knows that $s > 0$.

```
> int( exp(-s*(s+1)*t), t=0..infinity);
```

$$\frac{1}{\tilde{s}(\tilde{s}+1)}$$

Maple knows that $s > 0$, but we have not told it that $s > 1$, as we see in the following:

```
> int( exp(-s*(s-1)*t), t=0..infinity);
```

$$\lim_{t \to \infty-} -\frac{e^{(-\tilde{s}(\tilde{s}-1)t)}}{\tilde{s}(\tilde{s}-1)} + \frac{1}{\tilde{s}(\tilde{s}-1)}$$

```
> additionally(s>1);
> int( exp(-s*(s-1)*t), t=0..infinity);
```

$$\frac{1}{\tilde{s}(\tilde{s}-1)}$$

The above shows that Maple's built-in routines are capable, to some extent, of using assumed knowledge. This facility is still new to Maple and it has not yet propagated through all of Maple. You may wish to write your own programs to test knowledge of parameters: the routines to use are is and isgiven. See the help file for assume for details, and see also (17).

> "'No,'" she said, looking upon them daintily: "they have too
> many twiddles in them: like my Lord Lessingham's distich."'
> —E. R. Eddison, *Mistress of Mistresses*, p. 108.

Exercises

1. Assume $x > 0$, and verify that Maple knows then how to evaluate $\int_0^\infty \exp(-t)t^{x-1}\, dt$.

2. Assume that $s > 2$. Use Maple to compute the signum of s.

3. If x and y satisfy $x < -|y|$ and y is real, is the matrix

$$A = \begin{bmatrix} x & y \\ y & x \end{bmatrix}$$

negative semidefinite? See ?definite to get started, and use the Maple routine is. The answer to this exercise is contained in (17).

CHAPTER 2

Useful one-word commands

Maple has many built-in 'black boxes' for the simplification of algebraic expressions, solution of algebraic and calculus problems, standard manipulations from calculus, manipulations from the calculus of finite differences, solution of linear algebra problems, and evaluation of functions. This chapter examines some of the most useful of these black boxes.

2.1 Simplification

> '"That question," said Vandermast, "raiseth problems of high dubitation: a problem *de natura substantiarum*; a problem of selfness. Lieth not in man to resolve it, save so far as to peradventures, and by guess-work."'
> —E. R. Eddison, *A Fish Dinner in Memison*, p. 121

Simplification is, in general, an intractable problem. It is probably impossible to write a computer program that recognizes when arbitrary input expressions are equivalent to zero (21). This result, which is concerned with an infinite class of input expressions, translates into real difficulty in dealing with specific input expressions. For example, is $\log\tan(x/2 + \pi/4) - \sinh^{-1}\tan x = 0$? Certainly recognizing zero when you see it seems fundamental to simplification.

2.1.1 normal

However, if we restrict the class of functions we deal with, we can provide a *normal form* of representation. Zero is represented uniquely in a normal form: there are no nontrivial representations of zero in a normal form. For still further restricted classes of functions, we can provide a *canonical form* of representation. Each function is represented uniquely in a canonical form. Polynomials over the rationals have several possible canonical forms. For example, collect all like terms and sort them in ascending order. The Maple command `normal` puts multivariate rational polynomials with integer coefficients in a normal form, canceling common integer-coefficient factors by use of GCDs (21). This command should be used throughout computations with rational polynomials, since this keeps the size of the intermediate expressions down (generally speaking). For example,

```
> p := expand( (x+1)^3*(x+2)^2*(x+3) );
```

$$p := x^6 + 10\,x^5 + 40\,x^4 + 82\,x^3 + 91\,x^2 + 52\,x + 12$$

```
> q := diff(p, x);
```

$$q := 6\,x^5 + 50\,x^4 + 160\,x^3 + 246\,x^2 + 182\,x + 52$$

```
> r := p/q;
```

$$r := \frac{x^6 + 10\,x^5 + 40\,x^4 + 82\,x^3 + 91\,x^2 + 52\,x + 12}{6\,x^5 + 50\,x^4 + 160\,x^3 + 246\,x^2 + 182\,x + 52}$$

```
> normal(r);
```

$$\frac{1}{2}\,\frac{x^3 + 6\,x^2 + 11\,x + 6}{3\,x^2 + 13\,x + 13}$$

As an example showing that `normal` produces a normal form and not a canonical form, consider the following.

```
> Int(sqrt(tan(x)), x) = int(sqrt(tan(x)), x) + C;
```

$$\int \sqrt{\tan(x)}\,dx = \frac{1}{2}\,\sqrt{2}\arctan\left(\frac{\sqrt{2}\,\sqrt{\tan(x)}}{1 - \tan(x)}\right)$$
$$- \frac{1}{2}\,\sqrt{2}\ln\left(\frac{\tan(x) + \sqrt{2}\,\sqrt{\tan(x)} + 1}{\sqrt{1 + \tan(x)^2}}\right) + C$$

One way to test to see if Maple got the correct answer is to differentiate both sides and see if they give the same answer. This is a *necessary* condition, but not a sufficient condition: it is possible that Maple (or any other computer algebra system) will produce an answer that will pass this test but still be wrong. See section 2.3.1 for further discussion.

```
> diff(", x);
```

$$\sqrt{\tan(x)} = \frac{1}{2}\sqrt{2}$$

$$\left(\frac{1}{2} \frac{\sqrt{2}\left(1 + \tan(x)^2\right)}{\sqrt{\tan(x)}\left(1 - \tan(x)\right)} - \frac{\sqrt{2}\sqrt{\tan(x)}\left(-1 - \tan(x)^2\right)}{\left(1 - \tan(x)\right)^2} \right) \Bigg/$$

$$\left(1 + 2\frac{\tan(x)}{\left(1 - \tan(x)\right)^2} \right) - \frac{1}{2}\sqrt{2} \Bigg($$

$$\frac{1 + \tan(x)^2 + \dfrac{1}{2}\dfrac{\sqrt{2}\left(1 + \tan(x)^2\right)}{\sqrt{\tan(x)}}}{\sqrt{1 + \tan(x)^2}}$$

$$- \frac{\left(\tan(x) + \sqrt{2}\sqrt{\tan(x)} + 1\right)\tan(x)}{\sqrt{1 + \tan(x)^2}} \Bigg) \sqrt{1 + \tan(x)^2} \Bigg/ \Bigg($$

$$\tan(x) + \sqrt{2}\sqrt{\tan(x)} + 1 \Bigg)$$

Those expressions do *not* look equal. However, we can try to see if they each simplify to the same thing, as follows.

```
> normal(lhs("))=normal(rhs("));
```

$$\sqrt{\tan(x)} = \frac{1}{2}\frac{\sqrt{2}\left(\sqrt{2}\tan(x)^2 + 2\tan(x)^{3/2} + \sqrt{2}\tan(x)\right)}{\sqrt{\tan(x)}\left(\tan(x) + \sqrt{2}\sqrt{\tan(x)} + 1\right)}$$

Now, it is reasonably obvious to a human that both sides are equal, but the `normal` command did not simplify both expressions to the same form (i.e., `normal` does not produce a canonical form for expressions). It will, however, simplify the difference between these two expressions to zero.

```
> normal(lhs(") - rhs("));
```

$$0$$

Exercise

1. Plot Maple's answer to $\int \sqrt{\tan x}\, dx$, and show that it is *discontinuous* at $x = \pi/4$, well inside the domain of continuity of $\sqrt{\tan x}$. Conclude that Maple's antiderivative is only valid on a restricted domain, the size of which Maple does not tell you. See section 2.3.1 for more discussion of this difficulty.

2.1.2 collect

The Maple command `collect` can be used to put multivariate polynomials into a canonical form, with all coefficients of similar terms collected up. This function (or `expand`) must be called before the routines `coeff` or `coeffs` can pick off the coefficients. Another feature of `collect` is particularly useful in long computations—you can apply any function you like to each coefficient as the polynomial is collected.

```
> p := 1 + x + 3 + 5*x + 6*y + 17*y^2 + 35*x + 52*x^2 + 99*x*y + (x+y)^3;
```

$$p := 4 + 41\,x + 6\,y + 17\,y^2 + 52\,x^2 + 99\,x\,y + (\,x+y\,)^3$$

```
> collect(p, x);
```

$$x^3 + (\,52 + 3\,y\,)\,x^2 + (\,41 + 99\,y + 3\,y^2\,)\,x + 4 + 6\,y + 17\,y^2 + y^3$$

```
> collect(p, y);
```

$$y^3 + (\,17 + 3\,x\,)\,y^2 + (\,3\,x^2 + 99\,x + 6\,)\,y + 4 + 41\,x + x^3 + 52\,x^2$$

```
> collect(p, [x,y]);
```

$$x^3 + (\,52 + 3\,y\,)\,x^2 + (\,41 + 99\,y + 3\,y^2\,)\,x + 4 + 6\,y + 17\,y^2 + y^3$$

The command `collect(p, x, <function>)` applies the `<function>` function to each coefficient, once collected. Useful functions to use in this context include `factor` and `simplify`. The following more complicated example is discussed more fully in (12, vol. 2), but for now just observe the use of the applied function in `collect` to replace unwieldy expressions with more understandable labels.

```
> Weed_Index := 0:
> Homogeneous_Index := 0:  # Needed for general solutions.
> Computation_Sequence := array(1..10000):
> C := array(1..10000):  # 10000 ought to be enough.
```
The array C will be our 'placeholder' constants.

Remark on reuse of Maple-defined names. It turns out that the variable C has a special meaning in Maple: it refers to the routine for conversion of Maple expressions to C code fragments. One can still use this variable with relative impunity, because although assigning something to C removes any possibility of converting expressions to C code, that is the only Maple command that ceases to work (and indeed most users won't even notice that the command C no longer works in that session). See section 1.4.5.

To continue with the example of the use of collect, the routine Weed below will replace its argument with an unassigned constant from the array C, and remembers in the computation sequence what the actual argument was.

```
> Weed := proc(term)
>    local c: global Weed_Index, Computation_Sequence;
>    c := expand(term);      # Recognize zero if you see it.
>    if ( c <> 0) and nops(c) <> 1 and not type(c, numeric)
>       and not (type(c, '*') and type(op(1, c), constant)) then
>       Weed_Index := Weed_Index + 1:
>       Computation_Sequence[Weed_Index] := c:
>       C[Weed_Index]
>    else
>       c
>    fi
> end:
```

The procedure flatten first collects terms of like powers, then calls Weed to replace the coefficients with unevaluated constants, and records the values of these constants in the expression sequence Computation_Sequence.

```
> flatten := proc(expr, vars):
>    collect(expr, vars, distributed, Weed)
> end:
> p := randpoly([r, ln(r), a,b], dense);
```

$$p := -7 - 49\,r + 46\ln(r) - 61\,b - 71\,a + 56\,r^4\ln(r) - 46\,r^4\,a$$
$$- 95\,r^4\,b + 12\ln(r)\,a + 67\ln(r)\,b - 60\,a\,b + 40\,r^3\ln(r)^2$$
$$- 96\,r^3\ln(r) - 95\,r^3\,a^2 + 80\,r^3\,a + 80\,r^3\,b^2 + 67\,r^3\,b$$
$$- 59\ln(r)^2\,a + 43\ln(r)^2\,b + 41\ln(r)\,a^2 - 75\ln(r)\,b^2$$
$$- 80\,a^2\,b + 40\,a\,b^2 - 99\,r^2\ln(r)^3 + 59\,r^2\ln(r)^2 + r^2\ln(r)$$
$$- 94\,r^2\,a^3 + 81\,r^2\,a^2 - 54\,r^2\,a + 92\,r^5 + 33\,r^4 - 60\,r^3$$
$$- 94\ln(r)^2 + 32\,a^2 + 8\,b^2 + 98\,r^2 + 47\ln(r)^3$$
$$+ 32\,r^3\ln(r)\,a - 24\,r^3\ln(r)\,b - 17\,r^3\,a\,b - 46\ln(r)\,a\,b$$
$$- 85\,r^2\ln(r)^2\,a + 44\,r^2\ln(r)^2\,b - 13\,r^2\ln(r)\,a^2$$
$$+ 5\,r^2\ln(r)\,a - 57\,r^2\ln(r)\,b^2 - 72\,r^2\ln(r)\,b + 65\,r^2\,a^2\,b$$

$$+ 80\,r^2\,a\,b^2 - 19\,r^2\,a\,b - 37\ln(r)^2\,a\,b + 67\ln(r)\,a^2\,b$$
$$- 49\ln(r)\,a\,b^2 + 26\,a^3 + 93\,b^3 - 34\ln(r)^4 - 55\,a^4$$
$$+ 83\,r^2\ln(r)\,a\,b + 54\,r^2\,b^3 - 78\,r^2\,b^2 - 70\ln(r)^3\,a$$
$$- 75\ln(r)^3\,b - 53\ln(r)^2\,a^2 - 2\ln(r)^2\,b^2 - 86\ln(r)\,a^3$$
$$+ 23\ln(r)\,b^3 - 85\,a^3\,b - 53\,a^2\,b^2 - 63\,a\,b^3 - 22\,r\ln(r)^4$$
$$+ 83\,r\ln(r)^3 - 27\,r\ln(r)^2 - 79\,r\ln(r) - 95\,r\,a^4 + 54\,r\,a^3$$
$$+ 62\,r\,a^2 + 56\,r\,a - 30\,r\,b^4 - 41\,r\,b^3 - 5\,r\,b^2 - 17\,r\,b$$
$$+ 19\ln(r)^4\,a - 19\ln(r)^4\,b - 82\ln(r)^3\,a^2 - 87\ln(r)^3\,b^2$$
$$- 16\ln(r)^2\,a^3 - 35\ln(r)^2\,b^3 - 21\ln(r)\,a^4 - 71\ln(r)\,b^4$$
$$+ 83\,a^4\,b - 75\,a^3\,b^2 + 77\,b^4 + 33\,r\ln(r)^3\,a + 74\,r\ln(r)^3\,b$$
$$+ 13\,r\ln(r)^2\,a^2 - 77\,r\ln(r)^2\,a - 63\,r\ln(r)^2\,b^2$$
$$+ 26\,r\ln(r)^2\,b - 90\,r\ln(r)\,a^3 - 98\,r\ln(r)\,a^2 + 20\,r\ln(r)\,a$$
$$- 67\,r\ln(r)\,b^3 - 98\,r\ln(r)\,b^2 + 41\,r\ln(r)\,b - 41\,r\,a^3\,b$$
$$- 11\,r\,a^2\,b - 41\,r\,a\,b^3 - 84\,r\,a\,b^2 - 10\,r\,a\,b + 57\ln(r)^3\,a\,b$$
$$+ 38\ln(r)^2\,a^2\,b - 42\ln(r)^2\,a\,b^2 + 70\,r\,a^2\,b^2 - 79\ln(r)\,a^3\,b$$
$$- 72\ln(r)\,a^2\,b^2 - 65\ln(r)\,a\,b^3 - 26\ln(r)^5 - 62\,a^5$$
$$- 66\,r\ln(r)^2\,a\,b + 65\,r\ln(r)\,a^2\,b + 94\,r\ln(r)\,a\,b^2$$
$$- 48\,r\ln(r)\,a\,b + 79\,a^2\,b^3 - 21\,a\,b^4 - 30\,b^5 + 27\,r^2\,b$$

```
> flatten(p, [r,ln(r)]);
```

$$C_1 + C_2\,r + 92\,r^5 + C_3\,r^4 + C_4\,r^3 + C_5\,r^2 + 56\,r^4\ln(r) + 40\,r^3\ln(r)^2$$
$$+ C_6\,r^3\ln(r) - 99\,r^2\ln(r)^3 + C_7\,r^2\ln(r)^2 + C_8\,r^2\ln(r)$$
$$- 22\,r\ln(r)^4 + C_9\,r\ln(r)^3 + C_{10}\,r\ln(r)^2 + C_{11}\,r\ln(r)$$
$$+ C_{12}\ln(r) + C_{13}\ln(r)^2 + C_{14}\ln(r)^3 + C_{15}\ln(r)^4$$
$$- 26\ln(r)^5$$

The 'flattened' expression is obviously much more manageable, and perhaps is more understandable—the extraneous information contained in the constant terms has been hidden, removing some of the 'clutter' in the expression. Now, if we desire, we can examine the elements of the computation sequence:

```
> for i to Weed_Index do
> printf('C[%2i] = %a \n',i,Computation_Sequence[i]);
> od;
C[ 1] =  -7-80*a^2*b+40*a*b^2+79*a^2*b^3-21*a*b^4-61*b-71*a-75*a^3*b
C[ 2] =  -10*a*b-95*a^4+56*a-84*a*b^2-41*b^3+54*a^3-30*b^4+70*a^2*b^
C[ 3] =  -95*b-46*a+33
C[ 4] =  80*a-95*a^2-17*a*b+67*b-60+80*b^2
```

```
C[ 5] =  -78*b^2-94*a^3-19*a*b+27*b+80*a*b^2+54*b^3+98+65*a^2*b-54*a
C[ 6] =  32*a-96-24*b
C[ 7] =  59+44*b-85*a
C[ 8] =  5*a+83*a*b-72*b-57*b^2+1-13*a^2
C[ 9] =  83+33*a+74*b
C[10] =  -27-77*a-66*a*b-63*b^2+26*b+13*a^2
C[11] =  -98*a^2-79+94*a*b^2+41*b-98*b^2-48*a*b-90*a^3+65*a^2*b-67*b^
C[12] =  -46*a*b-86*a^3+23*b^3-79*a^3*b-21*a^4+67*b-75*b^2-65*a*b^3+1
C[13] =  -53*a^2-16*a^3-35*b^3+43*b-2*b^2-94-42*a*b^2-59*a+38*a^2*b-3
C[14] =  -75*b-82*a^2-87*b^2+47-70*a+57*a*b
C[15] =  19*a-34-19*b
```

I chopped this output at the right to fit it on the page.

2.1.3   factor

The Maple command `factor` is remarkably efficient at factoring multi-variate polynomials over the integers and occasionally over larger fields. It is also remarkable how often in applied problems nontrivial factorizations occur, and how useful it is to find them. For example, consider the following Maple session fragment. For simpler examples of factorization, see ?factor.

> `with(linalg):`

The following is MATLAB's gallery(3) matrix (see `help gallery` in MATLAB).

> `A := matrix( [[-149,-50,-154], [537,180,546], [-27,-9,-25]] );`

$$A := \begin{bmatrix} -149 & -50 & -154 \\ 537 & 180 & 546 \\ -27 & -9 & -25 \end{bmatrix}$$

> `e := matrix( [[130, -390, 0], [43, -129, 0], [133,-399,0]] );`

$$e := \begin{bmatrix} 130 & -390 & 0 \\ 43 & -129 & 0 \\ 133 & -399 & 0 \end{bmatrix}$$

> `At := evalm(A + t*e);`

$$At := \begin{bmatrix} -149 + 130\,t & -50 - 390\,t & -154 \\ 537 + 43\,t & 180 - 129\,t & 546 \\ -27 + 133\,t & -9 - 399\,t & -25 \end{bmatrix}$$

The characteristic polynomial of the above matrix is

```
> p := charpoly(At, x);
```

$$p := x^3 - 6x^2 + 11x - tx^2 + 492512tx - 6 - 1221271t$$

This has multiple roots when the discriminant (see (3)) is zero. So we choose t to make $d = 0$, where

```
> d := discrim(p, x);
```

$$d := 4 - 5910096t + 1403772863224t^2 - 477857003880091920t^3$$
$$+ 242563185060t^4$$

Let us compute a numerical value for the value of t that makes p have a multiple root. We can find an interval guaranteed to contain t by using realroot, as follows.

```
> readlib(realroot)(d, 1/10000000000);
```

$$\left[\left[\frac{13465}{17179869184}, \frac{6733}{8589934592}\right],\right.$$
$$\left.\left[\frac{8461218891559353}{4294967296}, \frac{33844875566237413}{17179869184}\right]\right]$$

So there are two roots, one small and one large.

```
> evalf(");
```

$$[[.7837661542\,10^{-6}, .7838243619\,10^{-6}],$$
$$[.1970031041\,10^7, .1970031041\,10^7]]$$

Thus we see that the small root is guaranteed to exist, and is (to three figures) $7.84 \cdot 10^{-7}$. This means that a small perturbation of the matrix A causes its eigenvalues (which are 1, 2, and 3) to collapse to one double and one single eigenvalue. This means that the eigenvalues of A are more difficult to compute than those of the average matrix. Unlike polynomial rootfinding problems, most eigenvalue problems are easy to solve numerically (44).

```
> alias(t_0=RootOf(d, t));
```

$$I, t_0$$

The variable t_0 is a symbolic way of representing *any* root of the discriminant. See (12, vol. 2) for a fuller discussion of RootOf.

```
> pt_0 := subs(t=t_0, p);
```

$$\mathrm{pt}_0 := x^3 - 6x^2 + 11x - t_0 x^2 + 492512 t_0 x - 6 - 1221271 t_0$$

Now pt_0 ought to have a multiple root.

> pf := factor(pt_0);

$$pf := 1/3617330840862075776271364437466707007583905327874048$$
$$(96703623305979008\,x + 19716665846146478987085\,t_0$$
$$- 28300556573899025 + 1103192925978112245349 5\,t_0{}^3$$
$$- 21733243079681277776111127375\,t_0{}^2)($$
$$-193407246611958016\,x + 19716762549769784966093\,t_0$$
$$+ 297216174096883795 + 1103192925978112245349 5\,t_0{}^3$$
$$- 21733243079681277776111127375\,t_0{}^2)^2$$

We see some application of Maple's large integers in the above factorization.

The routine `factor` can also factor over algebraic extensions of the integers. That is, if you know or suspect that the factors will contain, say, $\sqrt{2}$, you can ask Maple to factor into factors containing $\sqrt{2}$ by saying `factor(poly, sqrt(2))`. See `?factor` for more details.

2.1.4 expand

The Maple command `expand` is roughly the opposite of `combine`, which will be discussed in the next section. We have seen occasional examples of the use of expand in previous Maple sessions. More elementary examples can be found in the help file. The examples can be accessed directly by typing

> ???expand

and these examples show the ordinary use of `expand`. Using three question marks instead of one tells Maple that you wish to see only the entries under the EXAMPLES heading.

Of particular interest is the example that exhibits the so-called 'two-argument' form of `expand`:

> expand((x+1)*(y+z), x+1);

$$(x + 1)y + (x + 1)z$$

What this has done is expand the input expression $(x + 1)(y + z)$ while keeping the subexpression $x + 1$ unexpanded. This is *opposite* to the sense of the second argument to `combine` (see section 2.1.5) which tells combine to *act* on the terms containing the second argument: for expand, the second argument tells it what *not* to act on.

There is also a common use of expand not covered in ?expand, when expand is used with normal. Frequently this provides an essential part of result verification. The command normal usually tries to factor the numerator and denominator, and this can fail to provide a useful simplification. In that case, you pass the option expanded to normal, as follows.

```
> normal( 1/x + 1/x^2 - 1/(x+1) + 2/(x+1), expanded);
```

$$\frac{2\,x^2 + 2\,x + 1}{x^3 + x^2}$$

We used this in the first sample session, in 1.1.2.

2.1.5 combine

The Maple routine combine is a useful general-purpose routine for putting things together (see ?combine for examples). I find it most useful, however, for trigonometric simplification.

```
> cos(x)^3;
```

$$\cos^3(x)$$

```
> combine(", trig);
```

$$\frac{1}{4}\cos(3x) + \frac{3}{4}\cos(x)$$

This is particularly useful in perturbation calculations.

2.1.6 simplify

When all these alternatives fail or do not apply, one can try the command simplify. This routine applies various heuristics (that is, *ad hoc* techniques with little or no theoretical basis but that often work in practice) and can sometimes be invaluable. However, it is wise to keep a copy of the original object since sometimes simplify will make things worse, not better.

```
> e := cos(x)^5 + sin(x)^4 + 2*cos(x)^2 - 2*sin(x)^2 - cos(2*x):
> simplify(e);
```

$$\cos^5(x) + \cos^4(x)$$

The following example shows the use of `simplify` to prove something about continued fractions and Newton's method (33).

The continued fraction expansion for $\sqrt{2}$ is

$$1 + \cfrac{1}{2 + \cfrac{1}{2 + \cfrac{\cdot^{\cdot^{\cdot}}}{}}}$$

with convergents $c_0 = 1$, $c_1 = 1 + 1/2 = 3/2$, $c_2 = 1 + 1/(2 + 1/2) = 7/5$, $c_3 = 1 + 1/(2 + 1/(2 + 1/2)) = 17/12$, $c_4 = 41/29$, $c_5 = 99/70$, $c_6 = 239/169$, $c_7 = 577/408$, and so on. Newton's method applied to the function $f(x) = x^2 - 2$ with an initial guess $x_0 = 1$ produces iterates $x_1 = 3/2$, $x_2 = 17/12$, $x_3 = 577/408$, and so on, which apparently are all convergents of the continued fraction as well.

More generally, we can prove that $x_k = c_{2^k - 1}$ for the root of $f(x) = x^2 - Nx - 1$, using Maple for the algebra in the crucial step, as follows.

```
> Next := (x,N) ->  x - (x^2 - N*x - 1)/(2*x - N);
```

$$\text{Next} := (x, N) \to x - \frac{x^2 - Nx - 1}{2x - N}$$

That gives us the next iterate from Newton's method. One can prove that each convergent in the continued fraction will be of the form $\text{top}(k)/\text{bot}(k)$, where

```
> top := k ->  c*a^k + d*(-1/a)^k;
```

$$\text{top} := k \to c\,a^k + d\left(-\frac{1}{a}\right)^k$$

```
> bot := k ->  e*a^k + f*(-1/a)^k;
```

$$\text{bot} := k \to e\,a^k + f\left(-\frac{1}{a}\right)^k$$

where a is defined by

```
> N :=  a - 1/a;
```

$$N := a - \frac{1}{a}$$

We identify the constants c, d, e, and f by solving two linear systems.

```
> solve( {top(0) = N, top(1) = N^2 + 1}, {c,d});
```

$$\left\{ d = -\frac{1}{a\,(1 + a^2)}, c = \frac{a^3}{1 + a^2} \right\}$$

```
> assign(");
> solve( {bot(0) = 1, bot(1) = N}, {e,f});
```

$$\left\{ e = \frac{a^2}{1+a^2}, f = \frac{1}{1+a^2} \right\}$$

```
> assign(");
```

Now as an inductive step, suppose $X = \text{top}(k)/\text{bot}(k)$.

```
> X := top(k)/bot(k);
```

$$X := \frac{\dfrac{a^3\,a^k}{1+a^2} - \dfrac{\left(-\dfrac{1}{a}\right)^k}{a\,(1+a^2)}}{\dfrac{a^2\,a^k}{1+a^2} + \dfrac{\left(-\dfrac{1}{a}\right)^k}{1+a^2}}$$

The next iterate is

```
> Next(X, N):
> ans := simplify(");
```

$$\text{ans} := \left(-(-1)^{(2k)}\,a^{(4+2k)} + (-1)^{(3k)}\,a^{(4+4k)} + (-1)^{(2k)}\,a^{(2+2k)} - 2\,(-1)^{(3k)}\,a^{(6+4k)} \right.$$
$$+ (-1)^{(3k)}\,a^{(8+4k)} + 2\,a^{(4+2k)} + (-1)^{(3k)} - (-1)^k\,a^{(4+4k)} + a^{(6k+10)}$$
$$\left. + a^{(8+6k)} + a^2\,(-1)^{(3k)} - 3\,(-1)^{(1+k)}\,a^{(6+4k)} \right) \Big/ \left(a\,(1+a^2) \right.$$
$$\left(a^{(2+2k)} + (-1)^k \right)^2 \left(a^{(2+2k)} - (-1)^k \right) \Big)$$

Now test the above for equality with c_{2k+1}. The procedure `testeq` is a *probabilistic* equality tester—it gives a high degree of confidence that two expressions are equal by substituting strange (randomly chosen) values into each side.

```
> testeq(", top(2*k+1)/bot(2*k+1));
```

$$\text{true}$$

This result does not provide a *proof*, but the 'true' result is encouraging, and 'almost' a proof. We now use `normal` and `simplify` to show that they are really equal.

```
> normal(ans - top(2*k+1) / bot(2*k+1), expanded);
```

$$\left(-a^7 \, (a^k)^4 \, (-1)^k + a^7 \, (a^k)^4 \left((-1)^k\right)^3 - 2\,a^5 \, (a^k)^4 \left((-1)^k\right)^3 + 2\,a^5 \, (a^k)^4 \, (-1)^k \right.$$
$$- a^3 \, (a^k)^4 \, (-1)^k + a^3 \, (a^k)^4 \left((-1)^k\right)^3 - 2\,a^3 \, (a^k)^2 \left((-1)^k\right)^2 + 2\,a^3 \, (a^k)^2 \right) \Big/$$
$$\left(a^6 \, (a^k)^6 + a^4 \, (a^k)^4 \, (-1)^k - a^2 \, (a^k)^2 \left((-1)^k\right)^2 - \left((-1)^k\right)^3 + a^8 \, (a^k)^6 \right.$$
$$\left. + a^6 \, (a^k)^4 \, (-1)^k - \left((-1)^k\right)^2 a^4 \, (a^k)^2 - a^2 \left((-1)^k\right)^3 \right)$$

```
> simplify(");
```

$$0$$

This proves that the result of applying a Newton iteration to any convergent $P(k)/Q(k)$ of the continued fraction for the root of $x^2 - Nx - 1$ produces the convergent $P(2k+1)/Q(2k+1)$. The desired result about the quadratic convergence of Newton's method follows immediately by induction.

2.1.6.1 *Simplification with respect to side relations*

Sometimes you will want to simplify an expression subject to some constraints, or 'side relations'. One can do this in Maple by a special call to `simplify`, with a second argument consisting of a set of equations to be regarded as side relations. See `?simplify[siderels]` for examples.

Finally, sometimes `simplify` gets too enthusiastic and *over*simplifies things. See the exercises.

Exercises

1. Write down three or four reasonably complicated elementary functions, such as sin(exp(1/x)). Differentiate them using Maple, then integrate them again, and try to simplify the resulting answer to the original (plus some constant, of course).

2. Graph the functions $y_1 = \sin(\sin^{-1}(x))$ and $y_2 = \sin^{-1}(\sin(x))$ on $-10 \le x \le 10$. You should see graphically that y_2 is not always identically x. What does `simplify(arcsin(sin(x))` produce on your version of Maple? If it produces x, why is that wrong?

3. Give examples to show that neither the strategy of 'always expanding polynomials' nor the complementary strategy of 'always factoring polynomials' is optimal for simplification. That is, find examples of polynomials that are simpler in factored form, and find others that are simpler in expanded form.

4. Expand $(\cos\theta + i\sin\theta)^5$, and recover your input with `combine(", trig)`.

2.2 Solving equations

To solve algebraic equations or systems of algebraic equations, one can use the `solve` command.

```
> solve( x^3 - 1, x);
```

$$1, -\frac{1}{2} + \frac{1}{2}I\sqrt{3}, -\frac{1}{2} - \frac{1}{2}I\sqrt{3}$$

```
> solve(x^5-x+1, x);
```

$$\text{RootOf}(\_Z^5 - \_Z + 1)$$

```
> solve({x*t=1, x^2+t^2=4}, {x,t});
```

$$\left\{ x = \text{RootOf}(\_Z^4 + 1 - 4\_Z^2), \right.$$
$$\left. t = -\text{RootOf}(\_Z^4 + 1 - 4\_Z^2)^3 + 4\text{RootOf}(\_Z^4 + 1 - 4\_Z^2) \right\}$$

2.2.0.1 fsolve

To solve systems of equations numerically, one can use the `fsolve` (Float Solve) command.

```
> fsolve(x^5-x+1, x, complex);
```

$$-1.167303978, -.1812324445 - 1.083954101I, -.1812324445 + 1.083954101I,$$
$$.7648844336 - .3524715460I, .7648844336 + .3524715460I$$

```
> fsolve(x*tan(x)-1, x, 0..1);
```

$$.8603335890$$

This command can be used to find all complex roots of many polynomials, but complex roots of general functions are too much for it. Indeed, if the polynomial is at all difficult, `fsolve` will fail, and other methods are recommended.

2.2.0.2 *dsolve*

To solve differential equations, the routine `dsolve` is useful. It can solve many kinds of ordinary differential equations and initial-value problems analytically. It can also provide series solutions in many cases. See (12, vol. 2) for a discussion of series solution of differential equations.

```
> dsolve( {diff(x(t), t) = x(t) *(1 - x(t)), x(0)=alpha}, x(t));
```

$$x(t) = -\frac{1}{-1 - \dfrac{e^{-t}(1-\alpha)}{\alpha}}$$

Maple can also solve initial-value problems numerically. See `?dsolve, numeric` for details. Remember, however, that Maple is an interpreted language, and if your numerical problem is large or difficult you will probably be happier with a FORTRAN program such as those described in (26), available by anonymous ftp, or with the defect-controlled code such as PAMETH (18).

Maple does not currently have a routine for the numerical solution of boundary-value problems. I recommend instead the FORTRAN program COLNEW (2), available from `netlib`.

2.2.0.3 *rsolve*

One often wants to solve *finite-difference equations* (also known as *recurrence relations*). This is done in Maple with the `rsolve` command.

```
> eq := f(n) = f(n-1) + 2*f(n-2);
```

$$eq := f(n) = f(n-1) + 2f(n-2)$$

```
> rsolve({eq, f(0) = 1, f(1) = 1}, f(n));
```

$$\frac{1}{3}(-1)^n + \frac{2}{3}2^n$$

2.2.0.4   *linsolve*

Systems of linear algebraic equations can be solved many ways in Maple. For sparse systems of linear equations expressed in matrix form, the routine `linsolve` is efficient and simple.

```
> with(linalg):
> A := randmatrix(5, 5):
> b := vector(5, 1);
```

$$b := [\,1, \ 1, \ 1, \ 1, \ 1\,]$$

```
> x := linsolve(A, b);
```

$$x := \left[\frac{138417442}{10148464579}, \ \frac{-19420997}{922587689}, \ \frac{129221468}{10148464579}, \ \frac{4006336}{922587689}, \ \frac{169978120}{10148464579} \right]$$

```
> evalm(A&*x - b);
```

$$[0,0,0,0,0]$$

2.2.0.5   *Other solvers*

Solution of some Diophantine equations and some equations over finite fields is also possible in Maple. See `?isolve`, `?msolve`, and `?mod` for details.

2.2.1   Systems of polynomial equations

Systems of polynomial equations in more than one variable are much more difficult to solve, in general. Numerical techniques, such as homotopy methods (36), are often (perhaps even usually) much more effective than the best symbolic techniques, those of 'subresultants' and 'Gröbner bases' (21). Still, sometimes the symbolic techniques are what is wanted. Maple currently has implemented a heuristic substitution technique in its solve command, and a Gröbner basis method in its `grobner` package.

```
> eq1 := x^2 - y^2 - 1;
```

$$eq1 := x^2 - y^2 - 1$$

```
> eq2 := x^3 - 3*x*y^2 + 3*x^2*y - y^3 + 1;
```

$$eq2 := x^3 - 3\,x\,y^2 + 3\,x^2\,y - y^3 + 1$$

By doing this session twice, I noticed that to improve the presentation of the results I need the following alias.

```
> alias(alpha=RootOf(6*z+2*z^2+6*z^3+3, z));
```

$$I, \alpha$$

```
> solve({eq1, eq2}, {x, y});
```

$$\left\{ y = 0, x = -1 \right\}, \left\{ x = -\frac{3}{7}\,\alpha + \frac{6}{7}\,\alpha^2 + \frac{5}{7}, y = \alpha \right\}$$

The Gröbner basis technique 'reduces' the system of polynomials to a 'triangular' system of polynomials that is equivalent in that it has the same roots. The system is 'triangular' in that

1. the last equation is in only one variable,

2. the next-to-last contains only one new variable,

and so on. Thus if one has a reliable way to solve univariate polynomials one can solve the system written in terms of the Gröbner basis by solving a sequence of univariate polynomials. For this example, after solving the Gröbner basis we could see that it produced the same solutions as `solve` did. See (12, vol. 1) for a more thorough discussion of `RootOf`.

```
> with(grobner);
```

$$[\,\text{finduni, finite, gbasis, gsolve, leadmon, normalf, solvable, spoly}\,]$$

```
> gbasis({eq1, eq2}, {x, y}, plex, 'Y');
```

$$[\, 27\,y + 2\,y^2 + 7\,x + 7 + 24\,y^3, 6\,y^2 + 2\,y^3 + 6\,y^4 + 3\,y\,]$$

It is not clear which method is best for a given problem.

Remark. I remind the reader that the *numerical stability* of the resulting expressions (from `solve`) or the resulting system of polynomial equations (from `gbasis`) needs to be checked—just because you *can* make some particular mathematical transformation doesn't mean you *should* make that transformation. See the exercises.

Exercises

1. Solve $w^2 \exp w = x$ for w. See (12, vol. 2) or (14; 15) for more information about the Lambert W function, which appears in the solution.

2. Choose several random 2 by 2 polynomial systems of degree 2 and try to solve them with `solve` and `gsolve`. Use `time` to time them. Which is better?

3. Find approximate solutions to the systems of the previous question using `fsolve`.

4. Find the solution of $x^2 y'' + xy' + y = \sin(x)$, $y(1) = 1$, $y'(1) = 0$.

5. Find the solution of $p_n = 2np_{n-1} + n$ with $p_0 = 1$, for all integers $n > 0$.

6. Five sailors and a monkey are stranded on a desert island. They go out picking coconuts all day, and agree to split the coconuts evenly in the morning. After everyone else is asleep, one of the sailors divides the coconut pile into five equal piles, and finds that there is one left over, which he (quietly) splits and gives to the monkey. He then hides one pile for himself, and puts the other four piles back together. Satisfied, he goes to sleep. Another sailor then awakens, divides the pile into five and finds there is one left over, which he gives to the monkey. He, too, hides his pile and puts the rest back together. Similarly, each sailor does the same in turn. In the morning, when all awaken, they—with guilty grins all around, except for the monkey—split the pile into five, finding that it divides evenly into five piles, with one left over for the monkey. What is the minimum possible (positive) number of coconuts they could have started with? [Ans: 15621 coconuts (it was a busy day, evidently). Note that -4 coconuts would work if we didn't restrict the answers to be positive!]

7. Suppose A is a symmetric n-by-n matrix with distinct eigenvalues. Then it is known that its eigenvalues are *perfectly conditioned*, just because of the symmetry of the matrix (22). That is, small changes in the matrix will produce only small changes in the eigenvalues. Show by experiment that the Gröbner basis for the nonlinear system given by $Ax = \lambda x$ together with the normalization condition $||x||^2 = 1$ contains the characteristic polynomial of A (if the term ordering is taken with λ last). Since it is well known that most univariate polynomials are very poorly conditioned (44), conclude that computing a Gröbner basis can introduce spurious (and serious) numerical instability and thus may not be a practical way to calculate numerical roots of nonlinear systems.

2.3 Manipulations from calculus

Calculus is the algebra of limits, derivatives, integrals, and series, at least as far as Maple is concerned. The corresponding commands in Maple are `limit`, `diff`, `int`, and `series`. These black boxes are at the same time stronger and weaker than a good human calculator. They are stronger because they have more 'mathematical stamina' and can do longer calculations, and weaker because sometimes simple *ad hoc* techniques will give an answer where the standard techniques fail.

We begin with `diff`, the easiest of the routines to understand. There are several differentiation commands, including `diff`, `Diff`, `D`, the *automatic differentiation* (35) routine PD, and still others from the Maple Share Library. The command `Diff` is the 'inert' form, and doesn't actually *do* anything—delaying evaluation using inert forms is sometimes useful (see section 1.8.1). We will discuss D and other operators in a later chapter. See also the `linalg` routines `grad`, `jacobian`, and `hessian`.

2.3.0.1 *diff*

The routine `diff` differentiates expressions representing functions. It can find the derivatives of all elementary functions and many special functions. You can extend it so that it knows how to differentiate your own functions.

```
> diff(sin(x), x);
```

$$\cos(x)$$

```
> diff(exp(sqrt(x + y)), x);
```

$$\frac{1}{2} \frac{e^{\sqrt{x+y}}}{\sqrt{x+y}}$$

```
> diff(", y);
```

$$-\frac{1}{4} \frac{e^{\sqrt{x+y}}}{(x+y)^{3/2}} + \frac{1}{4} \frac{e^{\sqrt{x+y}}}{x+y}$$

```
> Diff(sin(x), x) = diff(sin(x), x);
```

$$\frac{\partial}{\partial x} \sin(x) = \cos(x)$$

```
> value(lhs("));
```

$$\cos(x)$$

See section 1.8.1 for a discussion of the command value.

What follows is a somewhat complicated example of extending the knowledge of Maple's diff routine. The following procedure tells Maple how to differentiate Chebyshev polynomials $T_k(x)$, represented in Maple as T(k,x). Note that the chain rule for differentiation is incorporated in this routine itself.

```
> 'diff/T' := proc(k, expr, x)
>    local j, ans;
>    if not type(k, 'integer') then
>       'diff(T(k, expr), x)'
>    elif k=0 then 0
>    elif k<0 then diff(T(-k, expr), x)
>    elif k=1 then T(0, expr)*diff(expr, x)
>    else
>       ans := - k*((-1)^(k-1)+1)/2*T(0, expr);
>       for j from 0 to trunc((k-1)/2) do
>          ans := ans + 2*k*T(k-1-2*j, expr);
>       od;
>       ans*diff(expr, x)
>    fi
> end:
```

For simpler examples of user-defined diff routines, see ?diff. The Chebyshev example above is not part of Maple already, and I felt it was useful to give a real but easily understood extension here. Note also the single quotes (which prevent evaluation) around the keyword integer. This provides some protection against the user typing something like integer := 18; before trying to differentiate $T_5(x)$, say.

```
> f := 3*T(0,x) + 4*T(1,x) + 7*T(2,x) + 1/4*T(3,x) + 8*T(4,x);
```

$$f := 3\,T(0,x) + 4\,T(1,x) + 7\,T(2,x) + \frac{1}{4}\,T(3,x) + 8\,T(4,x)$$

```
> df := diff(f, x);
```

$$df := \frac{19}{4}\,T(0,x) + 92\,T(1,x) + \frac{3}{2}\,T(2,x) + 64\,T(3,x)$$

```
> subs(T=orthopoly[T], f);
```

$$3 \, \mathrm{orthopoly}_T(\, 0, x\,) + 4 \, \mathrm{orthopoly}_T(\, 1, x\,) + 7 \, \mathrm{orthopoly}_T(\, 2, x\,)$$
$$+ \frac{1}{4} \, \mathrm{orthopoly}_T(\, 3, x\,) + 8 \, \mathrm{orthopoly}_T(\, 4, x\,)$$

```
> expand(");
```

$$4 + \frac{13}{4} x - 50 \, x^2 + x^3 + 64 \, x^4$$

```
> diff(", x);
```

$$\frac{13}{4} - 100 \, x + 3 \, x^2 + 256 \, x^3$$

```
> subs(T=orthopoly[T], df):
> expand("-"");
```

$$0$$

The above example showed some simple uses of `diff`, together with a nontrivial use of the user interface to `diff` to define the derivatives of the Chebyshev polynomials. See (12, vol. 2) for more details on this application. See `?orthopoly` for a description of the orthogonal polynomial package, which contains routines for expansion of certain orthogonal polynomials.

2.3.0.2   *int*

The next simplest command to use is `int`. This command will do both definite and indefinite integration of functions defined by expressions. The inert form `Int` provides a convenient interface to Maple's numerical quadrature routines. See section 1.8.1 for more details.

```
> int( sin(x), x);
```

$$- \cos(x)$$

Notice the conventional omission of the constant. If you want a constant, put it in yourself: `int(sin(x), x) + C` yields `-cos(x) + C`.

Maple can integrate functions first year students can't.

```
> int( ln(x)/(1+x), x);
```

$$\mathrm{dilog}(1 + x) + \ln(x) \ln(1 + x)$$

This integral is expressed in terms of the *dilogarithm function*. The dilogarithm function is

$$\text{dilog}(x) = \int_1^x \frac{\ln t}{1-t}\, dt .$$

This function often occurs in integrals arising from Feynman diagrams. Maple can do various computations with this function, including evaluating it for real values of x, plotting it, and computing series expansions. For example, here is what the share library package FPS (for Formal Power Series) can do with this function. See ?share for a discussion of the share library.

```
> with(share):
See ?share and ?share,contents for information about the share library

> readshare(FPS, calculus);
```

$$FormalPowerSeries$$

```
> ?FormalPowerSeries
> FormalPowerSeries(dilog(x), x=1);
```

$$-\left(\sum_{k=0}^{\infty} \frac{(-1)^k (-1+x)^{(k+1)}}{(k+1)^2} \right)$$

That series could also be presented as $\sum_{k=1}^{\infty}(1-x)^k/k^2$. This series converges if $0 < x < 2$.

Maple can also do *definite* integrals, as follows.

```
> int(exp(x^2), x=0..1);
```

$$-\frac{1}{2} I \sqrt{\pi}\, \text{erf}(I)$$

```
> evalf(");
```

$$1.462651746$$

Consider a call to Int (note the capital letter I at the beginning) instead of int.

```
> Int( exp(x^2), x=0..1);
```

$$\int_0^1 \exp(x^2)\, dx$$

Nothing has been done here: `Int` is *inert*. It is sometimes helpful to delay evaluation by using inert functions. See section 1.8.1.

```
> evalf(");
```

$$1.462651746$$

That invoked numerical integration. The numerical scheme is sophisticated and powerful, involving analysis of singularities and an arbitrary-precision technique selected from several available.

Maple can evaluate integrals containing parameters. This is a powerful feature, and allows things like Fourier series to be simply computed, as we saw in section 1.1.2.

```
> int(x^3*sin(m*x), x);
```

$$\frac{-m^3 x^3 \cos(mx) + 3 m^2 x^2 \sin(mx) - 6 \sin(mx) + 6 mx \cos(mx)}{m^4}$$

```
> collect(", [cos(m*x),sin(m*x)], expand);
```

$$\left(-\frac{x^3}{m} + 6\frac{x}{m^3}\right) \cos(mx) + \left(3\frac{x^2}{m^2} - 6\frac{1}{m^4}\right) \sin(mx)$$

2.3.0.3 *Unevaluated integrals*

"'My Lord Morville," replied Vandermast, "it is altogether a cross matter and in itself disagreeing, that you should expect from me an answer to such a question.'"
—E. R. Eddison, *A Fish Dinner in Memison*, p. 183

If `int` returns unevaluated, it means one of four things:

1. Maple has proved the integral is not elementary.

2. Maple has given up on the integral.

3. Maple knows how to do the integral for some special case of the values of the parameters, but is waiting for you to tell it if the parameters fall in that class. This changed in Maple V Release 2 to return an unevaluated limit in some cases.

4. Maple has, after computing the antiderivative, found a singularity in the interval and thinks the integral is undefined.

Here are some examples from each of the above categories.

```
> int(exp(x^4), x);
```

$$\int \exp(x^4)\, dx$$

This means the integral is not elementary. One can use `infolevel` to get the details of the proof from the Risch algorithm (21).

```
> infolevel[int] := 5;
```

$$\text{infolevel[int]} := 5$$

```
> int(exp(x^3), x);
int/indef:    first-stage indefinite integration
int/indef2:    second-stage indefinite integration
int/exp:   case of integrand containing exp
int/prpexp:    case ratpoly*exp(arg)
int/rischnorm:    enter Risch-Norman integrator
int/risch:    enter Risch integration
int/risch/algebraic1:    RootOfs should be algebraic numbers and functions
int/risch:    the field extensions are
```

$$[\_X,\ \exp(\_X^3)]$$

```
int/risch:    Introduce the namings:
```

$$\{\_th[1] = \exp(\_X^3)\}$$

```
unknown:    integrand is
```

$$\_th[1]$$

```
int/risch/exppoly:    integrating
```

$$\_th[1]$$

```
int/risch/diffeq:    solving Risch d.e.   y' + f y = g   where f,g are:
```

$$3\ \_X^2,\ 1$$

```
int/risch/DEratpoly:    solving Risch d.e.   y' + f y = g   where f,g are:
```

$$3\ \_X^2,\ 1$$

```
int/risch/exppoly:    Risch d.e. has no solution
int/risch:    exit Risch integration
```

$$\int \exp(x^3)\, dx$$

Here is an example from the second category, where Maple gives up.

```
> infolevel[int] := 0:
> int( 1/sqrt(a^4-x^4),x=0..a);
```

$$\int_0^a \frac{1}{\sqrt{a^4 - x^4}}\, dx$$

Maple ought to know how to do this integral, because it can do the case when $a = 1$, and as the following session shows, this is enough to let you do the general case.

```
> int(1/sqrt(1-x^4), x=0..1);
```

$$\frac{1}{4}\frac{\pi^{3/2}\sqrt{2}}{\Gamma\left(\frac{3}{4}\right)^2}$$

To show that this is not a problem from case 3, we issue the command

```
> assume(a>0);
```

and redo the computation.

```
> int( 1/sqrt(a^4-x^4), x=0..a);
```

$$\int_0^{a\tilde{}}\frac{1}{\sqrt{a\tilde{}^4-x^4}}\,dx$$

However, if we change variables, putting $x = au$,

```
> student[changevar](u=x/a, ",u);
```

$$\frac{1}{4}\frac{a\tilde{}\,\pi^{3/2}\sqrt{2}}{\sqrt{a\tilde{}^4}\,\Gamma\left(\frac{3}{4}\right)^2}$$

then Maple can get the answer. We can simplify it a bit, as well.

```
> simplify(");
```

$$\frac{1}{4}\frac{\pi^{3/2}\sqrt{2}}{a\tilde{}\,\Gamma\left(\frac{3}{4}\right)^2}$$

In case 3, some assumptions are needed before we can proceed.

```
> int(t^n, t=0..1);
```

$$\lim_{t\to 0+}\frac{t^{(n+1)}}{n+1}+\frac{1}{n+1}$$

The unevaluated limit here tells the user that assumptions on n must be made before Maple can continue with the analysis. Sometimes an uneval-uated integral means the same thing:

```
> int( exp(-t)*t^(x-1), t=0..infinity);
```

$$\int_0^\infty e^{-t}t^{x-1}\,dt$$

Here Maple knows how to do this integral but refuses to do so, because x might be negative.

Here is an example from the final category.

```
> int( 1/x, x=-1..1);
```

$$\int_{-1}^{1} \frac{dx}{x}$$

Maple refuses to do this one because the answer is undefined. Maple should return undefined. If you use the option CauchyPrincipalValue, you can force Maple to evaluate this integral.

```
> int( 1/x, x=-1..1, CauchyPrincipalValue);
```

$$0$$

See any complex variables text for a discussion of Cauchy Principal Value.

2.3.1 Continuity of antiderivatives

The following discusses a subtle bug in integration, present in most computer algebra systems. The bug is also present in many textbooks and tables so perhaps it is understandable why it persists. See (12, vol. 2) for a fuller discussion of the importance of being continuous.

```
> f := 1/(2+sin(x));
```

$$f := \frac{1}{2 + \sin(x)}$$

```
> F := int(f, x);
```

$$F := \frac{2}{3} \sqrt{3} \arctan\left(\frac{1}{6}\left(4\tan\left(\frac{1}{2}x\right) + 2\right)\sqrt{3}\right)$$

That integral is correct only in pieces. F is *discontinuous*, although there is a continuous antiderivative of f (as guaranteed by the fundamental theorem of calculus).

```
> plot(F, x=-3*3.14159..3*Pi, discont=true, colour=black);
```

This plot is shown in Figure 2.1, and we see clearly that this antiderivative is not satisfactory on any interval containing a discontinuity.

```
> limit(F, x=Pi, right);
```

$$-\frac{1}{3} \sqrt{3} \pi$$

Figure 2.1 Maple's antiderivative for $1/(2 + \sin x)$

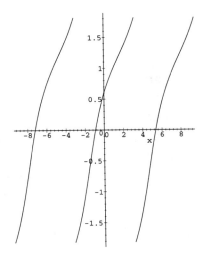

```
> limit(F, x=Pi, left);
```

$$\frac{1}{3}\sqrt{3}\,\pi$$

F has jump discontinuities at $x = n\pi$.

Maple uses some heuristics to correct this problem for a certain class of antiderivatives when it is asked to do a *definite* integral. The indefinite integral above is still unsatisfactory, though.

```
> int(f, x=-2*Pi..2*Pi);
```

$$\frac{4}{3}\sqrt{3}\,\pi$$

```
> evalf(");
```

$$7.255197458$$

```
> evalf(Int(f, x=-2*Pi..2*Pi));
```

$$7.255197457$$

Exercises

1. Find an antiderivative of $1/(2 + \sin x)$ that is continuous for all x. You may use the fact that $x/2 - \tan^{-1}(\tan(x/2))$ is piecewise constant. You may combine the arctangents by using the identity $\tan^{-1}(t_1) + \tan^{-1}(t_2) = \tan^{-1}((t_1 + t_2)/(1 - t_1 t_2))$.

2. Do the same for $1/(4 + \cos x)$.

3. Find a continuous antiderivative for $\sqrt{\tan x}$ by using Int and changevar to manually control Maple's integration.

4. If you have access to Derive, see if it gives continuous antiderivatives for the exercises in this section (my version gets two out of the first three).

5. Get Maple to solve $y'' + y + 1/4 \cos(3x) + 3/4 \cos(x)$, subject to $y(0) = 0$, $y'(0) = 0$. Show that Maple's solution is correct on $-\pi \le t \le \pi$ but not otherwise. The true solution is unbounded on $0 < t < \infty$ because of resonance. This bug may be corrected in a future version of Maple.

6. The time taken by a body to traverse part of its elliptical orbit around (say) the sun is proportional to the integral

$$\int_{\theta_0}^{\theta_1} r^2(\theta)\,d\theta$$

where the equation of the orbit is

$$r(\theta) = \frac{\text{constant}}{1 + \epsilon \cos \theta} \, .$$

Show that, for Maple, time does not progress at the 'stately pace of the planets' but instead does more of a 'hip-hop.'

2.3.1.1 limit

We have already seen examples of the use of limit, and no more needs to be said about it.

2.3.1.2 series

Series in Maple are not just Taylor series. Some examples showing the possible mathematical forms follow. See also section 3.3.

```
> series(sin(x^2), x,12);
```

$$x^2 - \frac{1}{6} x^6 + \frac{1}{120} x^{10} + O(x^{12})$$

```
> series(x^x, x,5);
```

$$1 - \ln\left(\frac{1}{x}\right) x + \frac{1}{2} \ln\left(\frac{1}{x}\right)^2 x^2 - \frac{1}{6} \ln\left(\frac{1}{x}\right)^3 x^3 + \frac{1}{24} \ln\left(\frac{1}{x}\right)^4 x^4 + O(x^5)$$

```
> series(x^x, x=1, 11);
```

$$1 + x - 1 + (x-1)^2 + \frac{1}{2}(x-1)^3 + \frac{1}{3}(x-1)^4 + \frac{1}{12}(x-1)^5$$
$$+ \frac{3}{40}(x-1)^6 - \frac{1}{120}(x-1)^7 + \frac{59}{2520}(x-1)^8 - \frac{71}{5040}(x-1)^9$$
$$+ \frac{131}{10080}(x-1)^{10} + O\left((x-1)^{11}\right)$$

```
> series(W(x), x);
```

$$x - x^2 + \frac{3}{2}x^3 - \frac{8}{3}x^4 + \frac{125}{24}x^5 + O(x^6)$$

The Maple routine asympt computes series about $x = \infty$. This asymptotic series computation can also be done with series(f, x=infinity).

```
> asympt(1/GAMMA(x), x);
```

$$\left(\frac{1}{2}\frac{\sqrt{2}}{\sqrt{\pi}\sqrt{\frac{1}{x}}} - \frac{1}{24}\frac{\sqrt{2}\sqrt{\frac{1}{x}}}{\sqrt{\pi}} + \frac{1}{576}\frac{\sqrt{2}\left(\frac{1}{x}\right)^{3/2}}{\sqrt{\pi}} + \frac{139}{103680}\frac{\sqrt{2}\left(\frac{1}{x}\right)^{5/2}}{\sqrt{\pi}}\right.$$
$$\left. - \frac{571}{4976640}\frac{\sqrt{2}\left(\frac{1}{x}\right)^{7/2}}{\sqrt{\pi}} - \frac{163879}{418037760}\frac{\sqrt{2}\left(\frac{1}{x}\right)^{9/2}}{\sqrt{\pi}} + O\left(\left(\frac{1}{x}\right)^{11/2}\right)\right)\left(\frac{1}{x}\right)^x e^x$$

```
> series(sin(x)^(1/3), x);
```

$$x^{1/3} - \frac{1}{18}x^{7/3} - \frac{1}{3240}x^{13/3} + O(x^{16/3})$$

```
> alias(alpha=RootOf(z^3+z+1, z));
```

$$I, \alpha$$

```
> series(RootOf(z^3+(1+x)*z+1, z), x);
```

$$\alpha + \left(\frac{2}{31}\alpha + \frac{9}{31}\alpha^2 + \frac{6}{31}\right)x + \left(\frac{25}{961}\alpha - \frac{27}{961}\alpha^2 - \frac{18}{961}\right)x^2$$

$$+ \left(\frac{67}{29791}\alpha - \frac{303}{29791}\alpha^2 - \frac{202}{29791}\right)x^3 + \left(-\frac{4114}{923521}\alpha + \frac{2598}{923521}\alpha^2 + \frac{1732}{923521}\right)x^4$$

$$+ \left(\frac{1927}{28629151}\alpha + \frac{51312}{28629151}\alpha^2 + \frac{34208}{28629151}\right)x^5 + O(x^6)$$

Reversion of series (14; 28) is also possible in Maple. For example, consider the function defined by $t \tan t = x$.

```
> Order := 11:
> series(t*tan(t), t);
```

$$t^2 + \frac{1}{3}t^4 + \frac{2}{15}t^6 + \frac{17}{315}t^8 + \frac{62}{2835}t^{10} + O(t^{12})$$

```
> solve(x=", t);
```

$$\sqrt{x} - \frac{1}{6}x^{3/2} + \frac{11}{360}x^{5/2} - \frac{17}{5040}x^{7/2} - \frac{281}{604800}x^{9/2} + O(x^{11/2})$$

The Maple routine `solve` uses series reversion techniques to do that. The above examples show that Maple series can be a little different from Taylor series. In particular, care must be taken to get a precise definition of what Maple means by its use of the O-symbol.

Definition: An asymptotic sequence $\{\phi_k(x)\}$ is a sequence of functions defined near a point $x = a$ (a might be ∞) such that each element of the sequence is smaller than the preceding ones. We say that one function f is smaller than g if $f = o(g)$ near $x = a$; that is, $\lim_{x \to a} f(x)/g(x) = 0$.

One simple definition of the O (big-oh) symbol is that $f = O(g)$ if there exists a nonzero constant K such that $\lim_{x \to a} f(x)/g(x) = K$. For simplicity we ignore here the possibility of oscillations, which require more refined treatment. Thus f and g are roughly the same size (up to a constant) near $x = a$. Maple allows the 'constant' K to vary, but not by much— if Maple says that f is $O(x^6)$ it means that there is a function $k(x)$ such that $x = o(k(x))$ and $f(x) = O(k(x)x^6)$ in the standard sense, or else that $x = o(1/k(x))$ and $f(x) = O(k(x)x^6)$. For example, look at the series for x^x computed in the previous example, and note that the coefficients of the powers of x contain terms of the form $\log^n x$. For any n, these vary more slowly than x near the point of expansion, $x = 0$. See `?series` and (9) for an alternate (but equivalent) description of the meaning of the O symbol in Maple.

When Maple computes a series, it ensures that each term is a member of an asymptotic sequence as above. The range of asymptotic sequences used by Maple is larger than most people normally use, and this explains the generalized series of Maple. However, there are many functions whose expansion requires asymptotic sequences not in the lexicon of Maple. The series package gdev by Bruno Salvy, available from the share library, improves on this, and recent work by Gaston Gonnet and Dominik Gruntz (23; 24) will improve it still further.

If you do not *want* such a generalized series, but rather want a strict Taylor series if it exists, then use the command taylor instead of series.

If you wish an *infinite* series, rather than the first few terms, the share library package FPS (for Formal Power Series) is useful. We saw an example of this before when we discussed the dilogarithm function.

Exercises

1. Tell Maple about the function $s(x)$, whose derivative is $s'(x) = \tan(s(x))$, by defining a procedure 'diff/s'. After that, ask Maple for the Taylor series of degree 3 for $s(x)$ about $x = 0$. This shows diff and series are linked. Use dsolve to find an explicit expression for $s(x)$ and verify that your series is correct.

2. Find an infinite series expansion for $y = \int_0^x \sqrt{1 + t^4}\, dt$.

3. The *exponential generating function* or 'egf' for the Bernoulli numbers is $t/(\exp(t) - 1)$. That is, the Bernoulli numbers B_k, $k = 0, 1, 2, \ldots$ appear in the series expansion

$$\frac{t}{e^t - 1} = \sum_{k=0}^{\infty} \frac{B_k}{k!} t^k \ .$$

Compute the first eight nonzero B_k in this way, and compare with the built-in bernoulli function.

4. The *ordinary* generating function or 'ogf' for the Catalan numbers is

$$F(x) = \frac{1 - \sqrt{1 - 4x}}{2x} = \sum_{k=0}^{\infty} f(k)x^k \ .$$

The difference between an ordinary generating function and an exponential generating function is the presence of $k!$ in the denominator of the terms from an exponential generating function. Compute the first eight Catalan numbers with series. Use FPS to find an explicit formula for all Catalan numbers.

5. Compute the first eight of the indicated polynomials by using their ordinary generating functions given below. Take the series with respect to t.

   (a) The Chebyshev polynomials of the first kind:
      ogf $= (1 - xt)/(1 - 2xt + t^2)$.

   (b) The Chebyshev polynomials of the second kind:
      ogf $= 1/(1 - 2xt + t^2)$.

   (c) The 'tree polynomials' (32): ogf $= 1/(1 + W(-t))^x$.

6. (From Bill Bauldry) Attempt to use Maple to compute the series for `max(x, 0)/x`. What should Maple return? Compare with Maple's answer for `series(signum(x), x)`.

2.4 Adding terms vs. the finite-difference calculus

Most of the time in Maple programming, when one wants to write an expression containing the sum of several similar terms, one does *not* want to use the routine sum, although sum can be (ab)used to do it. The following examples illustrate the proper constructs to accomplish this common task.

```
> s := convert( [seq(1/k, k=1..10)], '+');
```

$$s := \frac{7381}{2520}$$

Let us break that statement down into components. First, what does the seq command do?

```
> seq(1/k, k=1..10);
```

$$1, \frac{1}{2}, \frac{1}{3}, \frac{1}{4}, \frac{1}{5}, \frac{1}{6}, \frac{1}{7}, \frac{1}{8}, \frac{1}{9}, \frac{1}{10}$$

It creates an *expression sequence*. Now what are the square brackets for?

```
> ["];
```

$$\left[1, \frac{1}{2}, \frac{1}{3}, \frac{1}{4}, \frac{1}{5}, \frac{1}{6}, \frac{1}{7}, \frac{1}{8}, \frac{1}{9}, \frac{1}{10} \right]$$

They turn the sequence into a *list*. Now what does the `convert(<blah>, '+')` do?

```
> s := convert(", '+');
```

$$s := \frac{7381}{2520}$$

It simply added up the terms in the list.

An alternative way of doing the same thing uses a `for` loop. The above is slightly faster than a loop, and shorter to code, but is less efficient in memory, and is obviously more cryptic—I anticipate that a better way will be available in a future release of Maple, perhaps via a command called `add` with the same evaluation semantics as `seq`.

```
> s := 0;
```

$$s := 0$$

```
> for k to 10 do
>    s := s + 1/k;
> od:
> s;
```

$$\frac{7381}{2520}$$

The routine `sum` is really the finite-difference analogue of the routine `int`, and should be used only if you want a symbolic anti-difference (i.e., a sum of n terms expressed 'in closed form').

We clear the value of k since it was used before and has a value.

```
> k := 'k';
```

$$k := k$$

```
> sum(1/k, k=1..n);
```

$$\Psi(n+1) + \gamma$$

$\Psi(x) = \Gamma'(x)/\Gamma(x)$ is the derivative of the log of the Γ function. The routine `sum` had to do a reasonable amount of work to find that out, and if $n = 10$ the work was wasted. This is not serious for this example, because evaluation takes too little time to measure in this case, but the advantage of simple addition over symbolic summation grows quickly with the complexity of the symbolic problem.

```
> f := unapply(", n);
```

$$f := n \rightarrow \Psi(n+1) + \gamma$$

```
> f(10);
```

$$\frac{7381}{2520}$$

Note the use of unapply to create an operator or function from an expression. This is an extremely useful procedure and will be used over and over in this book. See ?unapply and ?operators for more details, as well as Chapter 3. See also the remarks on unapply in Section 1.1.2.4, for an explanation of the name 'unapply.'

Like Int, there is also an inert form Sum that is merely a 'place-holder' for the operation to be performed. The floating-point evaluation routine evalf knows about Sum and this provides a convenient user interface to floating-point evaluation of sums. However, evalf/Sum uses *Levin's u-transform* (43) to accelerate the convergence of a sum, and gives numerical values for convergent sums, slowly convergent sums, and even some divergent sums. Sometimes this is what is desired, as in the following example (this is pursued further in the chapter on the method of modified equations in (12, vol. 2)). If we define $c_1 = 1$ and

$$c_n = \frac{1}{1-n} \sum_{i=1}^{n-1} \binom{n-i+1}{i+1} c_{n-1}$$

where $\binom{n}{m} = n!/(m!(n-m)!)$ is the binomial coefficient, then the following program computes c_n for any n.

```
> c := proc(n) local i;
>    option remember;
>    -1/(n-1)*convert([seq(binomial(n-i+1, i+1)*c(n-i), i=1..n-1)], '+')
> end:
> c(1) := 1:
```

In the next chapter, we will be discussing option remember. In the procedure above, it makes a procedure for the evaluation of a recurrence relation more efficient (although the procedure could, with more programmer effort, be written more efficiently still *without* option remember). The manual assignment c(1) := 1 places the base of the recursion into the remember table. This particular procedure was given to me by Bruno Salvy, more as a convenient way to e-mail me the recurrence relation than anything else. See (12, vol. 2) for a discussion of the example from which this problem comes.

Note below the use of the inert Sum rather than sum to avoid needless symbolic processing here.

```
> Bseries := v -> evalf(Sum('c(n)*v^(n-1)', n=1..infinity));
```

$$Bseries := v \rightarrow evalf\left(\sum_{n=1}^{\infty} {}'c(n)\, v^{(n-1)\prime}\right)$$

```
> Bseries(0.05);
```

$$.9534459937$$

```
> Bseries(0.02);
```

$$.9805794612$$

That series is divergent, but nevertheless the results from evalf/Sum are correct (the function $B(v)$ can be defined by a convergent infinite product, and the values computed by the acceleration method agree with the values computed by the product).

Maple knows how to use symmetric functions to evaluate sums and products over the roots of polynomials. The following is a sum over all roots of a polynomial, done using rational means.

```
> alias(alpha=RootOf(z^6+z+1, z));
```

$$I, \alpha$$

```
> Sum(1/k, k=alpha);
```

$$\sum_{k=\alpha} \frac{1}{k}$$

```
> evalf(");
```

$$-1.000000000$$

Similarly for products.

```
> Product(1/(k+1), k=alpha);
```

$$\prod_{k=\alpha} \frac{1}{k+1}$$

```
> evalf(");
```

$$1.000000000 + .8241764125 \, 10^{-15} \, I$$

See (12, vol. 1) for more detail on this.

Exercises

1. Evaluate the following as 'closed form' functions of n.

   (a) $\sum_{k=0}^{n} \sin(k\pi/n)$

   (b) $\sum_{k=0}^{n} k(k-7)$ [Ans: $n(n+1)(n-10)/3$]

   (c) $\sum_{k=1}^{n} 1/k$ [Ans: $\gamma + \psi(n+1)$]

2. Evaluate the following using `evalf(Sum(...))`, correct to five decimal places. Ignore any occurrence of `FAIL` in the answer—it just means that an accuracy test was not passed, and Maple's 'best guess' at the answer is returned in brackets after the `FAIL`.

   (a) $\sum_{k=0}^{\infty} 1/(k^2+1)$ [Ans: 2.07667]

   (b) $\sum_{k=0}^{\infty} (-1)^k k! x^k$ for $x = 10$. Note the sum is divergent. Compare your answer to $\int_{t=0}^{\infty} \exp(-t)/(1+xt)\,dt$. Repeat for $x = 100$. [Ans: 0.201464 and 0.040785]

   (c) $\sum_{k=1}^{\infty} k^{-1/3}$. Again the sum is divergent. Since each term in the sum is positive, does Maple's answer make any sense? See section 1.8.1, and the definition of the Riemann ζ-function (e.g., see `?Zeta`). [Ans: $\zeta(1/3) \approx -0.973$]

3. Use Maple to add the following sums, but do not use `sum`.

   (a) $\sum_{k=0}^{100} (-1)^k x^k/k!$ for $x = 30$, in exact rationals and then using $x = 30.0$ instead so the calculations are done in floating-point arithmetic. Compare the answers. Repeat at higher settings of `Digits`, and explain the observed results. Note the infinite sum is convergent (and indeed the function is entire).

   (b) $\sum_{k=1}^{n} 1/k - \log(n+1/2)$ for $n = 100$, 1000, and 10000. Compare your answers with γ, the Euler-Mascheroni constant (see `?gamma`).

4. Evaluate $\prod_{k=\alpha} k/(2-k^3)$ where α satisfies $1 + \alpha + \alpha^2 + \cdots + \alpha^n = 0$ for $n = 10$, 20, and 40. Use `product` and exact arithmetic.

2.5 Floating-point evaluation

Arbitrary precision floating-point evaluation in computer algebra systems is overrated for its utility (this is heresy—or, at least, a controversial opinion). The philosophy behind arbitrary precision is that you attempt to buy more accuracy in your answer by spending more time and memory on precision in your calculations. This works sometimes, for simple calculations, but is seldom required in real applications. On the other hand, it is intellectually satisfying, and is occasionally really needed. Maple's `evalf` facility is a compromise between real efficiency for very large precisions and ease of programming, and is quite practical (as such things go).

There are two kinds of floating-point numbers used by Maple. The first is a Maple float, which is simply a pair of Maple integers wrapped in a call to the `Float` function: `Float(i,j)` means $i \cdot 10^j$, and prints in scientific notation. The second kind of floating-point number used by Maple is a 'hardware float,' which is either a floating-point number on a math co-processor or a C software float—so a Maple 'hardware float' might actually be a 'software float.' Maple floats are used by the `evalf` routine, and hardware floats are used by `evalhf`.

The Maple routine `evalf` is robust and reliable (in a sense to be discussed below). It will call `evalhf` (see below) if it thinks it can safely do so. Thus at low settings of `Digits` you get some of the speed benefits of hardware floating point. At higher settings of `Digits` the slower but more precise (and more accurate, since sometimes the chips get it wrong) software `evalf` subroutines come into play. If `Digits` > `evalhf(Digits)`, which shows the Maple syntax for discovering the approximate number of decimal digits a hardware float has (this changes, of course, with the system you are using Maple on), then `evalhf` certainly cannot be used. Maple will sometimes also use other information such as conditioning of the problem in making its decision to use `evalhf` on a given problem.

On a single binary operation or evaluation of one single built-in function, Maple's `evalf` routine claims 0.6 ulp (units in the last place) *relative* accuracy. That is, the Maple result is the exact result, rounded correctly to the number of decimal digits requested by the user. The 0.6, rather than the theoretically attainable 0.5, represents a reasonable compromise between attainable accuracy and efficiency.

No claim is made for the accuracy of more than one operation. That is to say, no claim is made for the accuracy of evaluation of an arbitrary *expression*, and indeed any such claim would have to be backed up by interval arithmetic or intelligence about the conditioning of the particular expression being evaluated.

Maple knows about the evaluation of several special mathematical constants, π, e, γ, and some others. To avoid wasting cycles on the computation of π Maple stores π to 10,000 places. Of course the time required to display 10,001 decimals of π is dramatically more! The other special constants are not stored to so many places (usually only 50 or so) as there is not so much interest in computing more digits of those constants.

The following examples illustrate the use and limitations of `evalf`. We begin with simple evaluation of constants.

```
> evalf(Pi, 50);
```

3.1415926535897932384626433832795028841971693993751

```
> evalf(gamma, 50);
```

.5772156649015328606065120900824024310421593359399 2

```
> evalf(E, 50);
```

2.7182818284590452353602874713526624977572470937

There are only 47 digits printed for *e* because trailing zeros were suppressed.

```
> evalf(Catalan, 50);
```

.9159655941772190150546035149323841107741493742816 7

Now examine evaluation of expressions.

```
> evalf(1 + sqrt(2) + sin(Pi/6) + cos(Pi^5) + x/400);
```

$$2.632481588 + .002500000000\, x$$

And now evaluation of special functions.

```
> evalf(erf(5), 30);
```

.999999999999846254020557196515 0

```
> Digits := 20;
```

$$\text{Digits} := 20$$

```
> evalf(BesselJ(0,10));
```

$$-.24593576445134833520$$

```
> evalf(LegendreF(1/2,1/sqrt(2)));
```

.53562273280540331970

Now a complex-valued example.

```
> evalf(Zeta(1/2+30*I));
```

$$-.12064228759004369991 - .58369121476370628876\, I$$

```
> evalf(GAMMA(0.2));
```

$$4.5908437119988030532$$

The routine `evalhf` is necessary for faster plotting, especially for complicated three-dimensional plots. If your machine has a floating-point co-processor, this routine attempts to use it (otherwise it just uses standard C doubles). This results in speed gains of approximately a factor of 30 over the `evalf` routine. Probably another factor of 40 would bring it into line with the numerical speeds of a language such as C or FORTRAN. If you wish to use `evalhf` in your programs, I offer the following tips:

1. Concentrate everything in one `evalhf`-able routine. Bringing hardware floats into Maple forces an automatic conversion to Maple floats, which usually negates the advantage of using hardware floats in the first place. You *can* pass arrays of hardware floats around inside *one* routine that has been passed to `evalhf`, from one subroutine to another. But any hardware floats that make it to the top level are converted.

2. Read the documentation carefully. Usually you will want to use `evalhf` to compute values in an array or vector, and you can do this; however, you will have to read and understand `?evalhf[arrays]` and `?evalhf[var]`.

3. Don't nest recursion too deep—15 levels seems to be the deepest you can go at the moment.

4. Do not use more than ten formal parameters in the procedure which you pass to `evalhf`. If you need more parameters in your function, pack them into an array.

5. Do not use more than 50 local variables in the procedure which you pass to `evalhf`. This is usually only a problem with automatically generated procedures, such as are produced by `optimize`.

The following example shows the use of `evalhf` to evaluate the terms in the Taylor series of the solution of the differential equation

$$\dot{y}_1 = 2y_1(1 - y_2)$$
$$\dot{y}_2 = y_2(1 - y_1)$$

about a given point $t = a$, where $y_1(a) = y_1^{(0)}$ and $y_2(a) = y_2^{(0)}$. This is problem B1 from the DETEST problem suite (30). [Incidentally, the exact solution for this problem can be expressed in terms of the Lambert W function (15).] The code for the Taylor series is easily derived from the Cauchy product formula, and requires $O(n^2)$ work to evaluate $O(n)$ terms in the Taylor series. The code was run using Maple V Release 3 on an IBM RISC machine.

```
ProblemB1 := proc(a, h, ya, n, y)
  local j,k,c,y1,y2;
  y[1,0] := ya[1];
  y[2,0] := ya[2];
  for k from 1 to n do
    c := 0;
    for j from 0 to k-1 do
      c := c + y[1,j]*y[2,k-j-1];
    od;
    y[1,k] := 2*(y[1,k-1] - c)/k;
    y[2,k] := (y[2,k-1] - c)/k;
  od;
end:
```

```
> ya := array(1..2, [1., 3.]):
```

```
> n := 200:
```

```
> y := array(1..2, 0..n):
```

```
> yf := array(1..2, 0..n):
```

```
> st := time(): evalhf(ProblemB1(0., 0.1, ya, n, var(y))): etime := time() -st;
```

$$etime := 1.070$$

```
> st := time(): evalf(ProblemB1(0., 0.1, ya, n, yf)): etimef := time() -st;
```

$$etimef := 9.540$$

```
> y[1,n];
```

$$.8164626396981414 \cdot 10^{61}$$

```
> yf[1,n];
```

$$.8164626512 \cdot 10^{61}$$

One sees in this example that use of evalhf resulted in a speed increase of a factor of 10, roughly. Note the use of var to tell evalhf that the routine ProblemB1 is expected to return an array of values.

2.6 The most helpful Maple utilities

My candidate for the most useful Maple command of all (aside, of course, from `help` or `?`) is the `read` statement. Without this to read in files of Maple commands, Maple would be almost unusable. Always use an editor to create files of Maple commands and the `read` statement to read them in. This saves hours of wasted effort retyping. The complementary output utilities are `writeto` and `appendto`. For example, the file `evf.mpl` contained the following statements:

```
# Simple evaluation of constants.
evalf(Pi, 50);
evalf(gamma, 50);
evalf(E, 50);
evalf(Catalan, 50);
# Evaluation of expressions.
evalf(1 + sqrt(2) + sin(Pi/6) + cos(Pi^5) + x/400);
# Evaluation of special functions.
evalf(erf(5), 30);
Digits := 20;
evalf(BesselJ(0, 10));
evalf(LegendreF(1/2, 1/sqrt(2)));
evalf(Zeta(1/2));
evalf(GAMMA(0.2));
```

To produce a preliminary version of the last example in the previous section, I started Maple, and then issued the command

```
> read 'evf.mpl';
```

which produced the output seen previously. I then used the `Print Session Log` menu item to save the output. Instead, I could have issued the commands

```
> writeto('evf.out');
```
```
> read 'evf.mpl';
```
```
> writeto(terminal);
```

which would have placed the output into the output file `evf.out`, together with the input commands, since I have set `interface(echo=2)` in my `maple.ini` file.

This 'batch-style' usage of Maple may disappear, when the use of worksheet-style Maple becomes the norm on all platforms. As previously stated, worksheets offer more convenience and better quality output, and in such an environment the use of Maple scripts as described above is less necessary.

The example in the previous section was actually created by using the *output* file from the `Print Session Log` command as input to Maple running under WIN/OS2. Importing output files in this way strips the results of the Maple commands and converts comments to text. I then re-executed the worksheet, after changing the Riemann ζ-function example to evaluate over the complex plane. I then exported the worksheet as a LaTeX input file, and used an editor to include that file in the source for this book.

This usage of worksheets as a 'final' pass on a file of input commands, to 'prettify' the output, is likely to remain a good way of using them for some time. Running your early files through the read command with the less-pretty output is considerably faster (at least on a small machine).

The routines readline and readstat read an expression ending in a semicolon from the terminal to use in a procedure—this is used if the procedure wants to ask the user a question. This style of usage is discouraged in Maple.

There are two new I/O utilities in Maple, sscanf and printf, based on the C utilities of the same names. These are useful for formatted I/O. See also readdata.

```
> for i to 10 do printf('i = %+2d and i^(1/2) = %+6.3f', i, evalf(sqrt(i))); od;
i = +1 and i^(1/2) = +1.000i = +2 and i^(1/2) = +1.414i = +3 and i^(1/2) = +1.73
```

I forgot the newline character \n in the format string. This is a common error. I truncated the line above so it would fit on the page. We now try again:

```
> for i to 10 do printf('i = %+2d and i^(1/2) = %+6.3f\n', i, evalf(sqrt(i))); od;
i = +1 and i^(1/2) = +1.000
i = +2 and i^(1/2) = +1.414
i = +3 and i^(1/2) = +1.732
i = +4 and i^(1/2) = +2.000
i = +5 and i^(1/2) = +2.236
i = +6 and i^(1/2) = +2.449
i = +7 and i^(1/2) = +2.646
i = +8 and i^(1/2) = +2.828
i = +9 and i^(1/2) = +3.000
i = +10 and i^(1/2) = +3.162
```

The +2d means a signed decimal integer in a field of width 2. Since +10 takes three characters it automatically widens the format (unlike FORTRAN which would have printed **).

```
> for i to 10 do printf('i = %+3d and i^(1/2) = %+6.3f\n', i, evalf(sqrt(i))); od;
i =  +1 and i^(1/2) = +1.000
i =  +2 and i^(1/2) = +1.414
i =  +3 and i^(1/2) = +1.732
i =  +4 and i^(1/2) = +2.000
i =  +5 and i^(1/2) = +2.236
i =  +6 and i^(1/2) = +2.449
i =  +7 and i^(1/2) = +2.646
i =  +8 and i^(1/2) = +2.828
i =  +9 and i^(1/2) = +3.000
i = +10 and i^(1/2) = +3.162
```

```
> for i to 10 do printf('i = % 3d and i^(1/2) = % 6.3f\n', i, evalf(sqrt(i))); od;
i =   1 and i^(1/2) =  1.000
i =   2 and i^(1/2) =  1.414
i =   3 and i^(1/2) =  1.732
i =   4 and i^(1/2) =  2.000
i =   5 and i^(1/2) =  2.236
i =   6 and i^(1/2) =  2.449
i =   7 and i^(1/2) =  2.646
i =   8 and i^(1/2) =  2.828
i =   9 and i^(1/2) =  3.000
i =  10 and i^(1/2) =  3.162
```

A blank instead of a + means positive numbers have a blank and negative numbers get a − sign.

See also `?system` for details on how to call other programs from within Maple.

The next most useful Maple commands are `alias` and `macro`. These allow you to use short names for convenience when the 'real' name of the object is quite long and awkward to type or read. I use `alias` most often together with `RootOf`, but it is also helpful in programming when you want to use long names to avoid conflicts with global variables but wish to use short names when typing. See (12, vol. 1) for further examples of `alias`. Note that it has been used throughout this chapter, and in particular in the last example of section 2.4.

The next two most useful Maple commands are arguably `map` and `unapply`. The first maps a function onto each element of some object (list, array, or set), and the second we have already seen used to create an operator from an expression. Similarly useful commands are `zip` and `select`.

```
> with(linalg):
> A := matrix(3, 3, (i,j) -> 1/(i+j+2));
```

$$A := \begin{bmatrix} \dfrac{1}{4} & \dfrac{1}{5} & \dfrac{1}{6} \\[2ex] \dfrac{1}{5} & \dfrac{1}{6} & \dfrac{1}{7} \\[2ex] \dfrac{1}{6} & \dfrac{1}{7} & \dfrac{1}{8} \end{bmatrix}$$

```
> evalf(A);
```

$$A$$

```
> map(evalf, A);
```

$$\begin{bmatrix} .2500000000 & .2000000000 & .1666666667 \\ .2000000000 & .1666666667 & .1428571429 \\ .1666666667 & .1428571429 & .1250000000 \end{bmatrix}$$

```
> s := [1,2,3,4,5,6];
```

$$s := [\,1,2,3,4,5,6\,]$$

```
> r := map( t -> 1/t^2, s);
```

$$r := \left[1, \frac{1}{4}, \frac{1}{9}, \frac{1}{16}, \frac{1}{25}, \frac{1}{36}\right]$$

```
> zip( (a,b)->[a,b], s, r);
```

$$\left[[1,1], \left[2, \frac{1}{4}\right], \left[3, \frac{1}{9}\right], \left[4, \frac{1}{16}\right], \left[5, \frac{1}{25}\right], \left[6, \frac{1}{36}\right]\right]$$

```
> select(has, ",1);
```

$$[[1,1]]$$

Now look at unapply.

```
> x^3 + 7*x^2*sin(1/x);
```

$$x^3 + 7\,x^2 \sin\left(\frac{1}{x}\right)$$

```
> f := unapply(", x);
```

$$f := x \rightarrow x^3 + 7\,x^2 \sin\left(\frac{1}{x}\right)$$

```
> f(-1);
```

$$-1 - 7\sin(1)$$

```
> f(-1.);
```

$$-6.890296894$$

The next most useful Maple commands are probably the code-generation routines fortran and C. They convert Maple expressions, vectors, or matrices into FORTRAN or C code. Both use a slightly counter-intuitive file interface in the function call instead of writeto or appendto. The Macrofort package, available from the share library, should be investigated, as it is a more serious package for code generation. The code optimization features of the built-in fortran and C, which use a 'janitorial' approach of trying to clean up an existing messy expression, are sometimes very useful.

```
> with(linalg):
> readlib(fortran):
> A := matrix(3, 3, (i,j) -> k^(i+j-1) );
```

$$A := \begin{bmatrix} k & k^2 & k^3 \\ k^2 & k^3 & k^4 \\ k^3 & k^4 & k^5 \end{bmatrix}$$

```
> fortran(A);
      A(1,1) = k
      A(1,2) = k**2
      A(1,3) = k**3
      A(2,1) = k**2
      A(2,2) = k**3
      A(2,3) = k**4
      A(3,1) = k**3
      A(3,2) = k**4
      A(3,3) = k**5
```

```
> fortran(A, optimized);
      t1 = k**2
      t2 = t1*k
      t3 = t1**2
      A(1,1) = k
      A(1,2) = t1
      A(1,3) = t2
      A(2,1) = t1
      A(2,2) = t2
      A(2,3) = t3
      A(3,1) = t2
      A(3,2) = t3
      A(3,3) = t3*k
```

The command fortran(A, optimized, filename='A.for') results in a file called A.for being created containing the same output as above. If the file already exists, the above output is appended.

```
> readlib(C):
> C(A);
      A[1][1] = k;
      A[1][2] = k*k;
      A[1][3] = k*k*k;
      A[2][1] = k*k;
      A[2][2] = k*k*k;
      A[2][3] = pow(k,4.0);
      A[3][1] = k*k*k;
```

```
        A[3] [2]  =  pow(k,4.0);
        A[3] [3]  =  pow(k,5.0);


>  C(A, optimized);
        t1  =  k*k;
        t2  =  t1*k;
        t3  =  t1*t1;
        A[1] [1]  =  k;
        A[1] [2]  =  t1;
        A[1] [3]  =  t2;
        A[2] [1]  =  t1;
        A[2] [2]  =  t2;
        A[2] [3]  =  t3;
        A[3] [1]  =  t2;
        A[3] [2]  =  t3;
        A[3] [3]  =  t3*k;
```

Exercises

1. Read the help file entries for `sscanf` and `readdata` and try them out.

2. Write a C or FORTRAN program that uses the cubic formula to find the roots of a given cubic equation. Use Maple to generate the cubic formula, and `C` or `fortran` to write the fragment of code at the heart of your program.

3. Use the `system` command to call some program external to Maple, perhaps a numerical program for solving a boundary value problem. Read the results back into Maple for further processing, perhaps plotting.

2.7   Plotting in Maple

'In that perfect hour all shadows had left earth and sky, and but form and colour remained: form, as a differing of colour from colour, rather than as a matter of line and edge (which indeed were departed with the shadows)...'
—E. R. Eddison, *A Fish Dinner in Memison*, p. 52

Maple is not a visualization language—it will not provide publication-quality graphics for you. However, its plotting facilities are improving (Release 3 plots are much better than previous versions) and are powerful enough to provide much insight; and its plots can be improved by draftists for publication if necessary. Almost 'raw' Maple plots are used in this book, however, since fidelity is more important in this context.

2.7.1 Two-dimensional plots

Maple has facilities for plotting graphs of functions represented by expressions, operators, or data; it can plot functions represented in Cartesian coordinates, polar coordinates, or parametrically. The relevant Maple command is `plot`.

As a simple example, consider using Maple to plot the Riemann ζ-function.

```
> plot(Zeta(t), t=-3..3, y=-3..3, discont=true, colour=black);
```

This graph was printed to an encapsulated PostScript file from WIN/OS2, and is shown in Figure 2.2.

Figure 2.2 The Riemann ζ-function

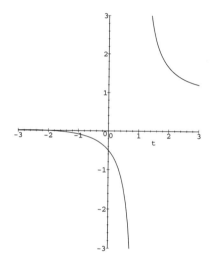

The option `discont=true` was used as a signal to Maple that the expression was discontinuous somewhere in the plot, and Maple then tried not to draw a line from the top to the bottom of the plot at the singularity (successfully in this case).

We can also put more than one plot on a graph, as follows.

```
> plot({W(x), W(-1, x)}, x=-0.5..1.5, y=-4..1, colour=black);
```

This plot, shown in Figure 2.3, shows the graph of the two real branches of $W(x)$ (see (12; 14; 15)) plotted on the same curve. There is some difficulty with plotting close to the branch point $(x, y) = (-1/e, -1)$, that shows up here as a 'hole' in the graph. Alternatively, we can plot this function parametrically, as follows.

```
> plot([t*exp(t), t, t=-4..1], x=-0.5..1.5, y=-4..1);
```

This parametric plot uses the definition of $W(x)$ as the number w such that $w \exp w = x$ to plot the graph completely (it is also much faster than

Figure 2.3 Lambert's W-function, drawn directly

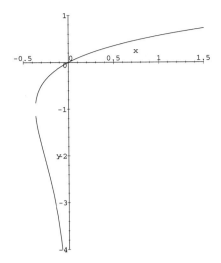

the previous plot, because $\exp x$ is a 'more built-in' function). This graph is shown in Figure 2.4.

Figure 2.4 Lambert's W-function, drawn parametrically

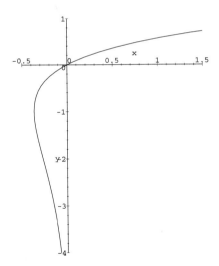

Now let us plot a collection of functions—say the Chebyshev polynomials.

```
> plot( { seq(orthopoly[T](k,x), k=0..20) }, x=-1..1, y=-1..1, colour=black );
```

This graph is shown in Figure 2.5.

Note the interesting curves that appear to be suggested by the places

Figure 2.5 The first 20 Chebyshev polynomials

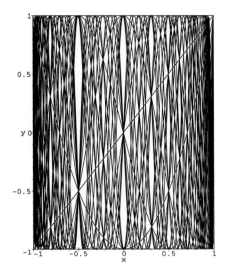

where the Chebyshev polynomials do *not* go. See (40) for a detailed explanation of these curves. The loci of these 'negative' curves are given by the equation

$$T_2(y) = T_q(x)$$

for $q = 1, 2, 3, \ldots$, and this can be solved parametrically by $y = \pm T_q(t)$, $x = T_2(t)$, because of the remarkable *composition property* of the Chebyshev polynomials, $T_m(T_n(x)) = T_n(T_m(x)) = T_{mn}(x)$ (see (40)). We plot the first few of these curves as follows.

```
> with(orthopoly):
> plot({seq([T(2,t), T(k,t), t=-1..1], k=1..5),
> seq([T(2,t), -T(k,t), t=-1..1], k=1..5)}, x=-1..1, y=-1..1, colour=black);
```

This graph is shown in Figure 2.6.

I cannot resist looking at the plot generated by the first 20 of these curves:

```
> plot( {seq([T(2,t), T(k,t), t=-1..1], k=1..20),
> seq([T(2,t), -T(k,t), t=-1..1], k=1..20)}, x=-1..1, y=-1..1, colour=black);
```

These curves, plotted in Figure 2.7, suggest that a similar analysis to that for the Chebyshev polynomials is possible here.

We may use the `plot` command indirectly, as with the following use of the `student` package.

```
> with(student):
> leftbox(1/(1+t), t=0..1, 6, colour=black);
```

This plot is shown in Figure 2.8.

Figure 2.6 The intersection loci

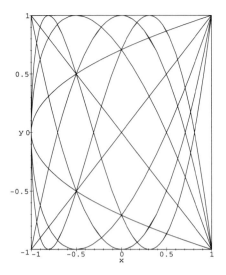

Figure 2.7 More intersection loci

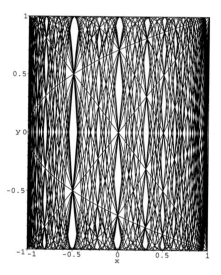

Now let us explore graphing partial sums of Fourier series. Compare this section with the sample session in section 1.1.2.

Consider this simple periodic function.

```
> f := 1/(2+sin(theta));
```

$$f := \frac{1}{2 + \sin(\theta)}$$

Figure 2.8 Riemann sums for
$1/(1 + t)$

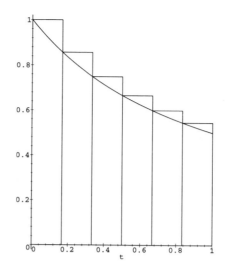

That function is periodic, and neither even nor odd. Hence its Fourier series
will contain both cosine and sine terms. We use the standard integrals to
compute a and b.

```
> restart;
> a := n -> Int( f*cos(n*theta), theta=0..2*Pi)/Pi;
```

$$a := n \rightarrow \frac{1}{\pi} \int_0^{2\pi} f \cos(n\,\theta)\, d\theta$$

```
> b := n -> (1/Pi)*Int(f*sin(n*theta), theta=0..2*Pi);
```

$$b := n \rightarrow \frac{1}{\pi} \int_0^{2\pi} f \sin(n\,\theta)\, d\theta$$

```
> Approx := N -> Sum( b(k)*sin(k*theta), k=1..N)
> + Sum(a(k)*cos(k*theta), k=1..N) + a(0)/2;
```

Approx :=

$$N \rightarrow \left(\sum_{k=1}^{N} b(k) \sin(k\,\theta) \right) + \left(\sum_{k=1}^{N} a(k) \cos(k\,\theta) \right) + \frac{1}{2} a(0)$$

```
> error := evalf(f-Approx(5)):
> plot(error, theta=0..2*Pi);
```

The plot is shown in Figure 2.9, and shows that the Fourier series gives a
good representation of the function.

Figure 2.9 The error in representing f by the first five terms in its Fourier series

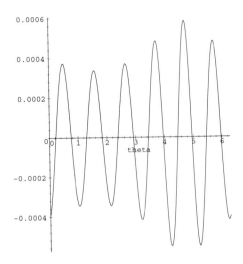

2.7.2 A highly discontinuous function

We now consider trying to graph the Gauss map $G : t \to t^{-1} \bmod 1$, from the theory of continued fractions. This function has discontinuities at $t = 1/n$, for all positive integers n. In Maple, this function can be defined as follows.

```
> G := t -> frac(1/t);
```

$$G := t \to \operatorname{frac}\left(\frac{1}{t}\right)$$

```
> plot(G, 0..1, 0..1, numpoints=101);
```

This plot is shown in Figure 2.10. It is a very ugly plot.

Now, perhaps that was unfair—there is a serious singularity at the origin, and we will have to deal with that ourselves.

What follows is an extended Maple session, that is intended for you to follow as if you are looking over my shoulder as I type. A significant amount of my own Maple knowledge was acquired by actually looking over the shoulder of various Maple experts—it's a good method. I will try to anticipate all your questions, but you can always try ? on any mysterious construct.

The session will provide an example of how to plot point-data in Maple, and more than one such plot on a graph. Let us choose 11 equally spaced points, in y, ranging from $1 - 10^{3-\text{Digits}}$ down to 0 (such a strange number was chosen only after some fiddling).

```
> y := seq( k/10., k=0..10);
```

Figure 2.10 First attempt to plot the Gauss map

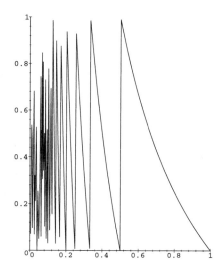

$$y := 0, .1000000000, .2000000000, .3000000000, .4000000000,$$
$$.5000000000, .6000000000, .7000000000, .8000000000,$$
$$.9000000000, 1.000000000$$

```
> y := [y]:
> y[11] := y[11]-Float(3,-Digits);
Error, cannot assign to a list
```

Oh, yes, I had forgotten about that. I should perhaps have used an array for y, but this is easily remedied, using subsop.

```
> y := subsop(11=1.-Float(1,3-Digits), y);
```

$$y := [0, .1000000000, .2000000000, .3000000000, .4000000000,$$
$$.5000000000, .6000000000, .7000000000, .8000000000,$$
$$.9000000000, .9999999]$$

This replaces the 11th operand in y with the desired quantity. Now let us specify the t-values that correspond to those y-values. There should be infinitely many, one on each piece of the graph, but we will make do with 100 pieces—which gives 1100 t-values.

```
> t := array(1..1100):
> for i to 100 do for j to 11 do t[(i-1)*11 + j] := 1./(i + y[j]); od: od:
> gt := map(G, t):
```

Let us look at the last few entries in that list, to see if the y-range spans the entire graph.

```
> gt_0 := map(G, [seq(t[k], k=1100-11..1100)]);
```

$$gt_0 := [.99999990, 0, .1000000, .2000000, .3000000, .4000000,$$
$$.5000000, .6000000, .7000000, .8000000, .9000000,$$
$$.9999999]$$

That shows that the map appeared to work—we now will get complete coverage of each interval. The following shows how to use `zip` to convert two separate lists of data points into a format acceptable for Maple's plotting facilities.

```
> points := zip((x,y)->[x,y], t, gt):
> plot(points, 0..1, 0..1);
```

Figure 2.11 An improved plot of
$G(t)$

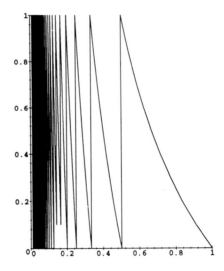

This plot is shown in Figure 2.11. It is better than before, but still not perfect. There are some gaps there (probably rounding errors), and lines jumping across the discontinuities. I want no extraneous lines. For some plots with discontinuities, you can use the option `discont=true` to tell Maple that the function is discontinuous and not to connect the graph across the discontinuities. However, this plot is too complicated for that approach to work. Instead, one way to fix this here is to create a whole bunch of plots, on the same graph.

```
> pieces := array(1..100);
> for i to 100 do
```

```
>    pieces[i] := NULL;
>    for j to 11 do
>    pieces[i] := pieces[i], 1./(i+y[j]), y[j];
>    od:
>    pieces[i] := [pieces[i]];
> od:
```

That program uses a loop to create an expression sequence, and then converts that expression sequence to a list. That construct is *inefficient*, because adding an entry to an expression sequence requires searching the whole expression sequence. Thus the above loop takes $1 + 2 + 3 + \cdots + n = n(n + 1)/2 = O(n^2)$ operations to create a list of length n. It is not crucial in this example because there are only 100 pieces, each of length 11, but it would have been better to use seq in the inner loop:

```
> pieces := array(1..100);
> ys := [seq(y[j], j=1..11)]:
> for i to 100 do
>    pieces[i] := zip((i,j)->(i,j), [seq(1./(i+y[j]), j=1..11)], ys);
> od:
```

This code is faster than the previous by a factor of about 5.

Now, to plot the pieces, we put them in a set, as follows.

```
> plot({seq(pieces[k], k=1..100)}, 0..1, 0..1, colour=black);
```

Figure 2.12 The best plot of $G(t)$

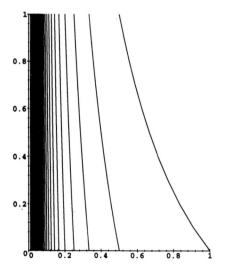

This plot is shown in Figure 2.12. Now *this* plot looks acceptable. There is still a little blank space next to the y-axis, because we have only plotted the pieces up to $n = 100$, and this leaves 1% of the graph uncovered. We will see yet another plot of this function, on a torus, in a subsequent section.

Remark. After having done that, it is clear that we could have done a similar job of plotting with much less effort, by simply re-parameterizing the curve into pieces: $(x, y) = (1/(n+t), t)$ as t runs from 0 to 1. Thus the command

```
> plot( {seq( [1/(n+t), t, t=0..1], n=1..100)}, x=0..1, y=0..1);
```

ought to produce nearly the same graph as before. We will not do this plot now, but rather reserve this idea for use with the torus plot later.

2.7.3   Polar coordinate plots

In (7), Michael W. Chamberlain draws the polar graphs of $r = \cos 5\theta + n \cos \theta$, for $0 \le \theta \le \pi$, for integers $n = -5$ (which gives a heart-shape) to $n = 5$ (which gives a bell-shape). We try to reproduce this in Maple.

```
> ?polarplot
> with(plots, polarplot);
```

$$[\, polarplot \,]$$

```
> polarplot({seq(cos(5*theta) + n*cos(theta), n=-5..5)}, theta=0..2*Pi);
```

That did *not* look like the picture in the College Mathematics Journal (this plot is not shown here). Possibly the problem is that Maple's `polarplot` facility disallows negative r-values. We try again using a parametric plot.

```
> r := (n,theta) -> cos(5*theta) + n*cos(theta);
```

$$r := (\, n, \theta \,) \to \cos(\, 5\,\theta \,) + n \cos(\, \theta \,)$$

```
> plot( {seq( [ r(n,t)*cos(t), r(n,t)*sin(t), t=0..Pi], n=-5..5)},
> x=-5..5, y=-4..4, colour=black);
```

The scale was not quite right on that plot, so we re-plot it with a different scale.

```
> with(plots, display):
```

There is a result returned by `with`, so the plot is now the *second*-last result.

```
> display("", view =[-5..6,-4..4]);
```

Now *that* plot looks like the one in the College Mathematics Journal. It is shown in Figure 2.13.

Figure 2.13 "Heart to Bell"

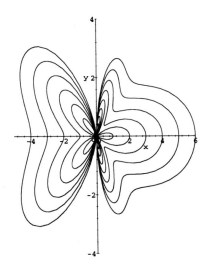

We now explore some other polar plots from an article by Fay (19).

```
> polarplot( (4*cos(3*theta) + cos(13*theta) )/cos(theta), theta=0..2*Pi);
```
See Figure 2.14.

Figure 2.14 $r = (4\cos 3\theta + \cos 13\theta)/\cos\theta$

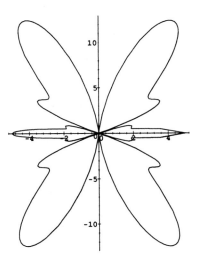

```
> polarplot( (4*cos(theta) + cos(9*theta) )/cos(theta), theta=0..2*Pi);
```
See Figure 2.15. Now try the *Fay butterfly*:

```
> polarplot( exp(cos(theta)) - 2*cos(4*theta) + sin(theta/12)^5,
>            theta=0..24*Pi );
```
See Figure 2.16. I selected the `axes=BOXED` option from the menu of the plot, rather than re-displaying the curve with a Maple command.

Figure 2.15 $r = (4\cos\theta + \cos 9\theta)/\cos\theta$

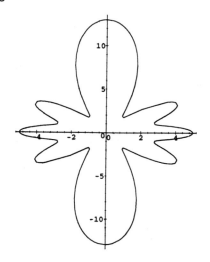

Figure 2.16 The Fay Butterfly
$r = \exp(\cos\theta) - 2\cos 4\theta + \sin^5(\theta/12)$

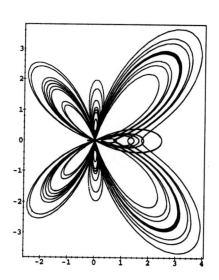

Exercises

1. Plot $1/\Gamma(x)$ on $-4 \le x \le 2$.
2. What will `plot(sin, -Pi/2..Pi/2)` produce?
3. Plot $\sin x$, $\cos x$, and $\tan x$ on the same graph, on $-4\pi \le x \le 4\pi$.
4. Plot the Folium of Descartes, which is given parametrically by
$$x = 3at/(1 + t^3), \qquad y = 3at^2/(1 + t^3).$$
   Nondimensionalize first, of course, and plot x/a *vs.* y/a.

5. Choose four random points in the x–y plane, fit a straight line to them (in the least squares sense), and plot the points and the line on the same graph. See `?regress` or `?leastsqrs`.

6. Plot $r = (4\cos m\theta + \cos n\theta)/\cos\theta$ for some odd values of m and n. If one of the values of m or n is even, the plot is supposed to look quite different. See (19) and the reference therein for more details.

7. Plot the Cissoid of Diocles, whose rectangular equation is

$$y^2 = \frac{x^3}{2a - x}$$

and whose parametric equations are $x = 2a\sin^2\theta$, $y = 2a\sin^3\theta/\cos\theta$.

8. Plot representatives of the Ovals of Cassini. The polar equation is $r^4 + a^4 - 2a^2 r^2 \cos 2\theta = b^4$. There are three qualitatively different curves, depending on whether b/a is less than 1, equal to 1, or greater than 1.

2.7.4 Three-dimensional plots

We continue with a plot of the Gauss map on a torus. The basic idea of this is that we want to take our flat graph, Figure 2.17, wrap it up in a tube, and then bend that tube around into a torus shape. Analytically, we are considering G as a map from the unit circle S^1 to itself; $G : S^1 \to S^1$ by $G(\exp(2\pi it)) = \exp(2\pi i/t)$, and now the 'fractional part' of the map is taken care of automatically. If we use the parameterization idea from the flat graph, what we get is the following.

First we define a torus, by specifying its centerline:

```
> sp := [rho*cos(2*Pi*t), rho*sin(2*Pi*t), 0, radius=b]:
```

Now let us define each piece of the curve (one 'wrap,' if you like), by

```
> pc := n -> [ (rho - r*cos(2*Pi*t))*cos(2*Pi/(n+t)),
>              (rho - r*cos(2*Pi*t))*sin(2*Pi/(n+t)),
>              -r*sin(2*Pi*t)]:
```

The sign of the last (z) component was chosen to make the graph agree with the cover of the March 1992 American Mathematical Monthly. Now we need to set suitable values for the parameters; in particular, we need to make the radius of the Gauss map slightly larger than the radius of the torus, so the hidden-line removal algorithms don't destroy it.

```
> rho := 3: r := 1.1: b := 1:
```

Now we do the plots by using routines from the `plots` package.

```
> with(plots):
```

The following generates the thickened curves, but does not display them. If we used a semicolon (;) we would see the plot structure line-printed, which is not what we want. The view is chosen (after some experimentation) to give a good scaling for the torus.

```
> s := spacecurve({seq(pc(k), k=1..50)}, t=0..1, thickness=2,
> colour=blue, view=[-4.4..4.4, -4.4..4.4, -2.2..2.2]):
```
This generates the torus defined with the centerline sp.
```
> s2 := tubeplot( sp, t=0..1, tubepoints=20,
> view=[-4.4..4.4, -4.4..4.4, -2.2..2.2]):
```
Now we display the two plots together.
```
> display({s, s2});
```

Figure 2.17 The Gauss map, graphed on a torus

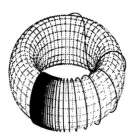

This code takes roughly 30 seconds to run under DOS on a 33MHz IBM clone, and roughly 2 minutes in WIN/OS2. After the plot has been displayed, one can use the menus to fiddle with the lighting schemes, etc., to produce the final version of the plot.

For another example, let us consider graphing $sn(u, k)$, which is one of the Jacobian elliptic functions. See (12, vol. 1) for a description of the Maple code to evaluate this function. This code should become available in the Maple Share Library some time in the near future.

```
> read 'ellifunt.m';
> with(ellifunt);
```

$$[\,cn, dn, sn, evalf/sncndn, init, set\_k\_to\,]$$

We plot $sn(u, k)$ on a compact x-interval because the convergence of $sn(u, k)$ to $\tanh(x)$ as $k \to 1^-$ is not uniform.

```
> plot3d(sn(x,y), x=-10..10, y=0..0.999999, grid=[30,30],
> style=HIDDEN, colour=black);
```
This graph is shown in Figure 2.18. Jon Borwein remarked that this is a 'graphical proof of the fast computability of the elementary functions,'

Figure 2.18 The Jacobian elliptic function $\mathrm{sn}(x, y)$

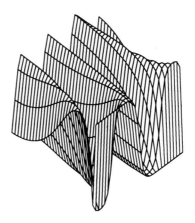

because it is known that the elliptic functions are quickly computable, and by continuation so is $\tanh(u)$.

The fact that $\mathrm{sn}(u, k)$ goes to $\tanh(u)$ is not really so evident in that graph—all the details and excitement of the limit happen for k close to 1. So we can expand that graph, by doing it logarithmically in k, as follows.

```
> plot3d(sn(x, 1-10^(-y)), x=-10..10, y=0..6, grid=[30,30]
> style=HIDDEN, colour=black, axes=BOXED);
```

Figure 2.19 shows the limiting case more clearly than the original figure.

Exercises

1. Use the `plot3d` command to plot $w = \Re(z^n)$ where $z = x + iy$ and $n = 3, 4, 5$, and 6. Rotate the plots and see if you can understand the pattern. Do the same for the imaginary parts of z^n.

2. Plot all the examples from `?plot3d`.

3. Plot the following functions.

   (a) $|x| + |y|$

   (b) $y^2/4 - x^2/9$

   (c) $(y^2 - x^2)/(y^2 + x^2)$

   (d) $\sin(\sqrt{1 - x^2 - y^2})/\sqrt{1 - x^2 - y^2}$

   (e) $(x^3 y - xy^3)/(x^2 + y^2)$

   (f) $e^{-y}\cos x$

Figure 2.19 The Jacobian elliptic function $\text{sn}(x, 1 - 10^{-y})$

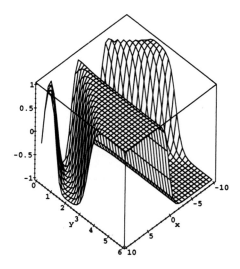

(g) $y^2 - y^4 - x^2$

(h) $1/(x^2 + 4y^2)$

(i) $xy^2/(x^2 + y^2)$

(j) $1 + \cos(x^2 + y^2)$

(k) $2xy/(x^2 + y^2)$

(l) $||x| - |y|| - |x| - |y|$

(m) $x = r\cos\theta, y = r\sin\theta, z = \cos(m\theta)/J_m(\lambda r)$, where $\lambda = 11.61984117$ and $J_m(z)$ is the Bessel J function of order m. Plot this for $m = 2$ and $m = 4$ on $0 \le r \le 0.9$ and $0 \le \theta \le 2\pi$. This plot was suggested by George Labahn.

2.7.5 Contour plots and other plots

We begin with some contour plots from (42). To produce contour plots in Maple, we first load the `plots` package.

```
> with(plots);
```

[animate, animate3d, conformal, contourplot, cylinderplot, densityplot, display, display3d, fieldplot, fieldplot3d, gradplot, gradplot3d, implicitplot, implicitplot3d, loglogplot, logplot, odeplot, pointplot, polarplot, polygonplot, polygonplot3d, polyhedraplot, replot, setoptions, setoptions3d, spacecurve, sparsematrixplot, sphereplot, surfdata, textplot, textplot3d, tubeplot]

As we can see, there are a good many other sorts of plots we can do with this package.

```
> f1 := sin(y-x^2-1) + cos(2*y^2-x);
```

$$f1 := -\sin(-y + x^2 + 1) + \cos(-2y^2 + x)$$

```
> contourplot(f1, x=-2..2, y=-2..2, grid=[100,100], colour=black);
```

Figure 2.20 Contours of $\sin(y - x^2 - 1) + \cos(2y^2 - x)$

That plot, shown in Figure 2.20, shows several local maxima, minima, and saddle points. We will give only one more contour plot here directly, but will suggest several other interesting plots in the exercises. Consider now

```
> f2 := y + sin(x^2*y-1/x);
```

$$f2 := y + \sin\left(x^2 y - \frac{1}{x}\right)$$

Actually, Skala (42) considers $f = \sin(y + \sin(x^2 y - 1/x))$. Initially I thought that the contours of the above function should be the same as those of Skala's function, because if $\sin(f) = $ constant, then $f = $ constant also—hence the shape of each contour will be the same. However, the

Figure 2.21 Contours of $y +$ $\sin(x^2 y - 1/x)$

appearance of the two contour plots is quite different. Why this is so is left to the exercises.

```
> contourplot(f2, x=-Pi..Pi, y=-Pi..Pi, grid=[100,100], colour=black);
```
This plot is shown in Figure 2.21.

Exercises

1. Read the documentation for the `plots` package.

2. Plot the contours of $f = \sin(y + \sin(x^2 y - 1/x))$ and discuss why it looks different from Figure 2.21.

3. Plot the contours of

   (a) $w = \Re(z^n)$, where $z = x + iy$ and $n = 3, 4, 5,$ and 6.

   (b) $1/(x^2 + y^2 - \pi) + \exp(x + y/\pi)$ on some suitable range around the origin.

   (c) $x^4 + y^4 - 6x^2 y^2$

   (d) $(x - y)/(x + y)$

   (e) $\tan(x^2 + y)$

   (f) $\sin(x^2 - y^2 - 1) + \cos(4x^2 y^2)$

   (g) $xy + 2x - \ln(x^2 y)$

   (h) $x = r \cos \theta$, $y = r \sin \theta$, and $z = \cos(m\theta)/J_m(\lambda r)$, where $\lambda = 11.61984117$, $J_m(z)$ is the Bessel J-function of order m. Choose $m = 4$ and $m = 2$. This plot was suggested by George Labahn.

4. Plot the graph of the surface given implicitly by $x^3 - y^2 + z^3 = 1$. (See `?implicitplot3d`.)

We can find some plotting utilities especially designed for ordinary differential equations in the `DEtools` package.

```
> with(DEtools);
```

$$[\text{DEplot}, \text{DEplot1}, \text{DEplot2}, \text{Dchangevar}, \text{PDEplot}, \text{dfieldplot},$$
$$\text{phaseportrait}]$$

The following shows how to plot a phase diagram for the Van der Pol equation

$$\ddot{x} - \epsilon(1 - x^2)\dot{x} + x = 0 \,,$$

using the usual first-order system form that arises on putting $y = \dot{x}$. We choose $\epsilon = 1$ and 5. Other phase portraits are left for the exercises.

```
> phaseportrait([y, (1-x^2)*y-x], [x,y], 0..40, {[0,2.,0]}, x=-3..3, y=-3..3,
> stepsize=0.05);
```

Figure 2.22 Phase portrait for the Van der Pol equation, $\epsilon = 1$

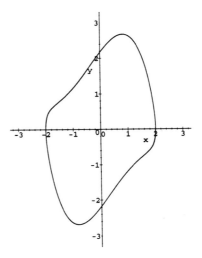

```
> phaseportrait([y, 5*(1-x^2)*y-x], [x,y], 0..40, {[0,2,0]}, x=-3..3, y=-8..8,
> stepsize=0.05);
```

We now look at a plot of the solution to a first-order differential equation, $y' = \cos(ty)$, for various initial conditions. For an explanation of the curious 'bunching' of the curves, see the exercises in (4). The `seq` construct in the following plot command merely specifies 19 different initial conditions, and hence we expect to see 19 different curves in the plot. The `stepsize` specification is to ensure that enough points are generated to give

Figure 2.23 Phase portrait for the Van der Pol equation, $\epsilon = 5$

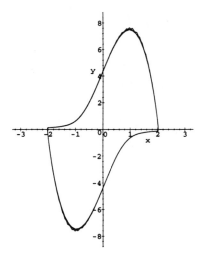

good graphical accuracy. This is inefficient for more complicated systems, since it requires more function evaluations than interpolation would; however, for larger systems one would want to use purely numerical packages anyway.

```
> DEplot1(cos(t*y), [t,y], t=0..6, {seq([0,k/3], k=0..18)}, stepsize=0.1,
> y=0..6, arrows=NONE);
```

Figure 2.24 Solution of $y' = \cos(ty)$ for various initial conditions

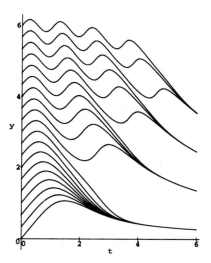

Now let us examine a second-order differential equation system, the Lotka-Volterra predator-prey equations.

```
> DEplot2([-3*x+4*x^2 -x*y/2 - x^3, -2.1*y+x*y], [x,y], 0..10,
> {[0,-1,-1], [0,-1,0.3], [0,-1,1.5], [0,-0.75,4], [0,0.25,4], [0,4,3.7],
> [0,4,2.6], [0,4,2], [0,4,1], [0,4,0.25], [0,4,-0.75], [0,3,-1], [0,1,-1]},
> x=-1..4, y=-1..4, stepsize=0.1, arrows=NONE);
Over/underflow in conversion to doubleOver or underflow in conversion to do
```

The error message (which I truncated so it would fit on the page) is a bit bothersome—one doesn't know if the results can be trusted. However, the plot shown in Figure 2.25 agrees with ref. 38 (Fig. 1.1, p. 7), with other reliable numerical solutions of this equation, and with the linear stability analysis of the fixed points, and so no harm was done in this case.

Figure 2.25 Phase plane solutions to predator-prey equations

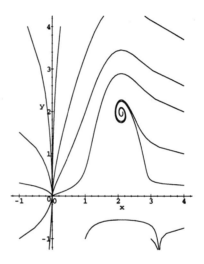

Exercises

1. Plot a dodecahedron.

2. Plot $y = \int_0^x 1/\Gamma(t)\,dt$ on $-4 \le x \le 2$, first by direct use of plot and then by plotting the solution of the differential equation $y' = 1/\Gamma(x)$ using DEplot1. Which is faster?

3. Plot the curve defined by $x^5 + y^5 = 5ax^2y^2$, first by using implicitplot (nondimensionalize, of course, so you don't have to deal with the a), and then by putting $y = tx$ and deriving a pair of parametric equations, $x = f(t)$ and $y = tf(t)$.

4. The continuous logistic equation is $\dot{x} = x(1-x)$, $x(0) = \alpha$. Solve this analytically in Maple and plot the solution for $\alpha = 0, 1/5, 2/5, \ldots, 9/5, 2$. Compare the time it takes to do this with the time taken to generate the plot by use of DEplot1.

5. The discrete logistic equation is $x_{n+1} = \mu x_n(1 - x_n)$, $x_0 = \alpha$. It arises, for example, by applying Euler's method to the continuous logistic equation, but also directly in applications. Write a Maple program to approximate the attractive periodic points of this map, given the numerical value of μ. For values of μ varying between 0 and 4, plot these periodic points.

6. Learn how to use `textplot` and `textplot3d` to annotate your plots.

7. Use `conformal` to examine the complex transformations $w = \ln(z)$ and $w = (z^2 - 1)^{1/2}$.

8. Plot a pentagon and a five-pointed star on the same graph. See `polygonplot`, and use $[\cos(2\pi k/5), \sin(2\pi k/5)]$ to generate the co-ordinates of the vertices of the pentagon and a similar sequence for the five-pointed star.

2.7.6   Common errors

Probably the single most common error is *believing what you see*. Computer-generated plots can be misleading, even if the software is doing what it is supposed to. Sometimes the misleading behaviour is obvious (plot, for example, $\sin(x^2)$ on $0 \le x \le 30$, or $\sin(1/x)$ on any interval containing 0). At other times, it is not obvious. Plot

$$x^7 - \frac{7x^6}{2} + \frac{161x^5}{32} - \frac{245x^4}{64} + \frac{6769x^3}{4096} - \frac{3283x^2}{8192} + \frac{3267x}{65536} - \frac{315}{131072}$$

on $-1 \le x \le 2$ and note that it looks completely flat on $0 \le x \le 1$. However, there are seven real roots equally spaced on this interval, and thus six extrema. This fact is not at all evident from the graph, on this scale. In fact, automatic scaling of polynomials so that interesting features can be seen is a difficult problem, and I have not yet seen a program that can do it satisfactorily.

Common Maple mistakes include:

1. Forgetting that x has a value.

   ```
   > x := 17:   # Now a big, long session, so we forget about x.
   > plot(sin(x), x=0..5);
   Error, (in plot/options2d) unknown or bad argument, 17 = 0 .. 5
   ```

   To fix this problem, unassign x by x := 'x'; and reissue the plot command.

2. Plotting a function defined by a procedure that expects only numerical arguments without preventing premature evaluation. Suppose for

example we wished to plot a piecewise-defined function defined by the
following.

```
> p := proc(t);
>    if t > 1 then
>       sin(Pi*t) + 1
>    else
>       t^2
>    fi
> end:
> plot(p(x), x=0..2);
Error, (in p) cannot evaluate boolean
```

This error message arises because p has been evaluated at the symbolic
argument x, and Maple can't tell if $x > 1$ or not. One way to fix it is
to use quotes to prevent premature evaluation.

```
> plot('p(x)', 'x'=0..2);
```

Another way is to plot using the operator syntax. This technique is
highly recommended.

```
> plot(p, 0..2);
```

Still another way is to rewrite the procedure to return unevaluated if
the argument is not numeric.

```
> p := proc(t);
>    if not type(t, numeric) then
>       'p'(t)
>    elif t > 1 then
>       sin(Pi*t) + 1
>    else
>       t^2
>    fi
> end:
> plot(p(x), x=0..2);
```

3. Trying to plot functions that contain symbols is impossible. For example, one cannot plot $r = a \cos \theta$ in polar coordinates. One should
   nondimensionalize and plot r/a vs. θ instead.

2.7.7 Getting hardcopy of your plots

There are many different printer styles supported by Maple. For details,
see ?interface[plotoutput] or ?plotsetup. Printing from a windowing

session is usually as easy as finding the correct menu item—that is how most of the graphs were generated for this book.

Here we show how to directly produce a PostScript plot.

The following function $B(v)$ arises in the analysis of the effect of solving $y' = y^2$ by Euler's method. This is pursued further in (12, vol. 2). For now, we note that $B(v)$ satisfies a functional equation

$$B(v) = \frac{(1+v)^2}{1+2v}B(v+v^2)$$

and has a series expansion beginning

$$B(v) = 1 - v + \frac{3}{2}v^2 + O(v^3)\,.$$

We have already seen the Maple code for generating an arbitrary number of the coefficients for this series, in section 2.4. It is reproduced here for convenience.

```
> c := proc(n) local i;
>     option remember;
>     -1/(n-1)*convert([seq(binomial(n-i+1,i+1)*c(n-i), i=1..n-1)], '+')
> end:
> c(1) := 1:
```

The following uses Levin's u transform to accelerate the sum, which turns the asymptotic series into a convergent one. This only works, by experiment, if $-0.1 \le v \le 0.1$.

```
> Bseries := v -> evalf(Sum('c(n)*v^(n-1)', n=1..infinity));
```

$$\text{Bseries} := v \rightarrow \text{evalf}\left(\sum_{n=1}^{\infty} {}'\text{c}(\,n\,)\,v^{(n-1)\prime}\right)$$

We use two infinite product representations to fill in the rest of the values of B.

```
> B := proc(v) local p,v0,u0;
>     if not type(v, numeric) then
>        'B(v)'
>     elif v < -1 then
>        (1.+v)^2/(1.+2*v)*B(v*(v+1.))
>     elif v=-1 then
>        0.
>     elif -1 < v and v < 0 then
>        v0 := v;
>        p  := 1.;
>        while v0 < -0.1 do
>           p := p*(1.+v0)^2/(1+2.*v0);
>           v0 := v0*(v0+1.);
>        od:
```

```
>      p*Bseries(v0)
>    elif v = 0 then
>      1.
>    else
>      u0  := v;
>      p   := 1.;
>      while u0 > 0.1 do
>        u0 :=  2.*u0/(1.+(1.+4.*u0)^(1/2));
>        p  := p*(1.+2.*u0)/(1+u0)^2;
>      od:
>      p*Bseries(u0)
>   fi
> end:
```

At last, the commands to actually produce the plot.

```
> plotsetup(ps, plotoutput='beyn.ps', plotoptions='portrait, noborder');
```

```
> plot(B, -2..2, -2..2);
```

This function is difficult to graph in any lower-level language.

The above session produces in the file 'beyn.ps' the following PostScript commands. Sending the file to a PostScript printer produces the plot shown in Figure 2.26.

```
%!PS-Adobe-3.0 EPSF
%%Title:
%%Creator: MapleV
%%Pages:  1
%%BoundingBox: 0 0 612 792
%%DocumentNeededResources: font Courier
%%EndComments
20 dict begin
gsave
/m {moveto} def
/l {lineto} def
<****** many PostScript commands omitted ******>
(1) dup stringbbox 3290 exch sub exch 2 idiv 3500 exch sub exch m show
(2) dup stringbbox 3290 exch sub exch 2 idiv 4500 exch sub exch m show

showpage
grestore
endo
%%Trailer
%%EOF
```

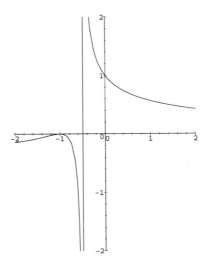

Figure 2.26 The graph of $B(v)$

CHAPTER

3 Programming in Maple

> '[Lady Fiorinda was] in the theoric
> of these matters liberally grounded through daily sage expositions
> and informations by Doctor Vandermast, who had these four
> years past been to her for instructor and tutor. To try her paces
> and put in practice the doctor's principles and her own most
> will-o'-the-wisp and unexperimental embroiderings
> upon them, ready means lay to hand...'
> —E. R. Eddison, *A Fish Dinner in Memison*, p. 137

Maple is useful as a collection of 'black boxes,' but it is more useful still as a very high-level programming language. Since most of the tasks undertaken by Maple are 'one-off' calculations (as opposed to 'batch' calculations, which require many executions of the same program), it makes sense that Maple is an interpreted, rather than compiled, language. This is true even for the Maple library, because for large problems the cost will be dominated by the manipulation of large objects. Some crucial operations, though, are performed by kernel routines, which *are* compiled for efficiency.

Maple procedures can be divided loosely into two types: operators and more general procedures. An operator is meant to imitate a mathematical operator, both in notation (insofar as this is possible in ASCII) and in actions. The first section of this chapter deals with operators and their uses. More general procedures, studied in a later section of this chapter, can do essentially anything computable. Since Maple is a high-level language you can express these actions in many ways.

For more in-depth information on how to program in Maple, see (34) which is available from the Maple share library. For an extended example of revising a program for efficiency, see (41). For examples of useful programs, see (20) and (12).

3.1 Operators

Operators are functions from one abstract space to another. The mathematical usage of the word *operator* is usually reserved for functions on functions. For example, consider the differentiation operator $D : f \rightarrow f'$. The Maple notation for this operator is just D. Historically D was introduced into Maple to solve one particular problem: to find a way to represent the value of the derivative of an unknown function at a particular point (the natural command subs(x=3, diff(f(x), x)) doesn't work—it produces diff(f(3),3) which gives an error). This is represented in Maple now as D(f)(a), meaning $f'(a)$. Note that Newton's notations $\dot{x}(a)$ and $x'(a)$ survive today because they denote this idea concisely; contrariwise, the Leibniz notation df/dx survives because it has its own algebraic and mnemonic uses. The notation

$$\frac{df}{dx}\bigg|_{x=a}$$

is relatively cumbersome. I know of only one computer algebra system (that of the HP48 series calculators) that uses this notation.

Operators in Maple are represented by 'arrow' notation:

```
> f := t -> t*sin(t);
```

$$f := t \rightarrow t \sin(t)$$

or, alternately, by 'angle-bracket' notation f := <t*sin(t) | t>, which is useful if you wish to define a 'local variable' in your operator (see section 3.4). Angle-bracket notation may be removed from Maple shortly, however, if the arrow syntax can be coerced into taking local variables in a nice way. In contrast, the t in the above operator is an *argument*, and not, strictly speaking, a local variable.

```
> f(0);
```

$$0$$

```
> f(1.5);
```

$$1.496242480$$

```
> f(x^2+a);
```

$$(x^2 + a) \sin(x^2 + a)$$

```
> D(f);
```

$$t \rightarrow \sin(t) + t \cos(t)$$

Those were examples of function *application*. We *applied* the operator f to the arguments 0, 1.5, and $x^2 + a$ in turn. That result, $(x^2 + a) \sin(x^2 + a)$, is the result of *composing* the operator f with the operator $g : t \mapsto t^2 + a$, and *applying* the resulting operator to the argument x. *Application is not the same as composition.*

3.1.0.1 *Application* vs. *composition*

The following rules of thumb do not give the whole story, but they help.

- When you compose two operators, the result is an operator. The domain and range of the two operators must be compatible.

- When you apply an operator to an argument, the result is an expression. The argument must be in the domain of the operator.

Mathematically, this is a difference in *what you think about the answer*, and is subject to a little deliberate overlap when this is useful. In Maple the distinction is usually a bit more strictly enjoined by syntax.

The Maple syntax for function application uses parentheses ( ). The function f applied to the argument u is written `f(u)`.

The distinction between application and composition will be made clearer with further examples, but note for now that the result of composing f with g is the operator $f \circ g : t \mapsto f(g(t))$, whereas the application of this operator to the argument x gives the *expression* $f(g(x))$. To be concrete, if $f = t \rightarrow \sin(t + \phi)$ and $g = u \rightarrow \exp(u)$, then $f \circ g = z \rightarrow \sin(\exp(z) + \phi)$ is an operator, while $f(g(x)) = \sin(\exp(x) + \phi)$ is an expression, because we think of x as being a real number.

3.1.0.2 *More operator examples*

The following is one definition of an antidifferentiation operator. Typically, this is denoted I, but that symbol is taken in Maple.

```
> Aye := f -> unapply(int(f(x), x), x);
```

$$\text{Aye} := f \rightarrow \text{unapply} \left(\int f(x)\, dx, x \right)$$

```
> Aye(f);    # f from before
```

$$x \to \sin(x) - x\cos(x)$$

Some simplifications of operators are automatic:

```
> f := t -> sin(t);
```

$$f := \sin$$

That is just the sin operator.

```
> f := t -> exp(t);
```

$$f := \exp$$

That is just the exp operator.

Operators can sometimes arise on output, as in this series computation.

```
> f := 'f':
> series(f(x), x=a);
```

$$f(a) + D(f)(a)(x-a) + \frac{1}{2}D^{(2)}(f)(a)(x-a)^2 + \frac{1}{6}D^{(3)}(f)(a)(x-a)^3$$
$$+ \frac{1}{24}D^{(4)}(f)(a)(x-a)^4 + \frac{1}{120}D^{(5)}(f)(a)(x-a)^5 + O\left((x-a)^6\right)$$

Again, application is not the same as composition. Here we have applied the operator D to the operator f, which yields another operator $D(f)$. This is best thought of as application, not composition, because the operator f is *in the domain of* D, and thus is best thought of as an *argument*. If we wish to *compose* two operators, their domains and ranges must be compatible.

We then *apply* the resulting operator $D(f)$ to a to get $D(f)(a)$ or $f'(a)$. We *compose* D with itself, and apply the resulting operator to f to get $D \circ D(f) = D^{(2)}(f)$, and apply this resulting operator to the argument a to get $D^{(2)}(f)(a) = f''(a)$. If we mix these two ideas, we generate errors in Maple.

```
> D(g(h));
Error, (in D) univariate operand expected
```

The composition operator in Maple is the at-sign @. This is the closest symbol available to the standard mathematical notation ∘.

```
> D(g@h);
```

$$D(g)@h \, D(h)$$

```
> D(g*h);
```

$$D(g)h + gD(h)$$

Repeated composition in Maple uses the @@ operator. Thus $D^{(4)}$ is represented in Maple as D@@4.

Maple's D operator knows both the chain rule and the product rule. If you know h is constant and x is the independent variable, you can use the remember table of D to simplify some computations:

```
> D(h) := 0;
```

$$D(h) := 0$$

```
> D(x) := 1;  # This says that dx/dx = 1.
```

$$D(x) := 1$$

```
> D(F@x);
```

$$D(F)@x$$

The following exhibits an inconsistency that may change with a future version of Maple.

```
> D(x^2);  # Remember, we have stated D(x) = 1
```

$$2x$$

```
> D(g^2);
```

$$2D(g)g$$

```
> D(x^3+h);  # Again, we have stated D(x) = 1 and D(h) = 0
```

$$3x^2$$

```
> D(g^3 + H);
```

$$3D(g)g^2 + D(H)$$

```
> D(sin(x));
Error, (in D) univariate operand expected

> D(sin(g));
Error, (in D) univariate operand expected

> D(sin@g);
```

$$\cos@g \, D(\, g\,)$$

```
> D(sin@x);  # Again, remember that we stated D(x) = 1
```

$$\cos@x$$

To my way of thinking, $D(\sin(f))$ should be interpreted as differentiation of the composition of the operator sin with the operator f. This is what happens if the 'outer function' is a simple power, as we saw: g^2 is treated as the composition of the squaring operator with the operator g, and $D(g^2)$ simplifies to $2gD(g)$. Since we told Maple that $D(x) = 1$, $D(x^2)$ reduced to $2x$. It appears that only the integer powering functions get this special treatment at present.

3.1.1 Finite-difference operator examples

Operators can act on functions (other operators) or on numerical objects. As an example, we investigate the finite-difference operators. We begin with the shift operator.

```
> Eh := <unapply(f(x+h), x) | f | x >:
```

The above is the preferred construction for this shift operator. It uses a global variable h for the shift, but a local variable x in its construction (to avoid difficulties with any possible previously defined global value for x). An equivalent way of doing this would be

```
> Eh := f -> unapply(f(DUMMY+h), DUMMY):
```

using a global variable DUMMY that hopefully no one else has used, and yet another construction with the same purpose would be

```
> Eh := f -> subs('DUMMY'=f, (x->DUMMY(x+h))):
```

The subs of the dummy global variable DUMMY is necessary because this version of Maple has no nested lexical scopes. So we must use a temporary global variable and substitute it for the applied argument. Both of these latter constructions are inferior to the first. The use of quotes allows this procedure to work even if someone has assigned a value to DUMMY.

```
> Eh(sin);
```

$$x \to \sin(x + h)$$

```
> Eh(cos);
```

$$x \to \cos(x + h)$$

```
> Eh(exp);
```

$$x \to e^{(x+h)}$$

Now the shift operator is related to the finite difference operator Δ by

```
> Delta := f -> Eh(f) - f;
```

$$\Delta := f \to \mathrm{Eh}(f) - f$$

```
> Delta(sin);
```

$$(x \to \sin(x + h)) - \sin$$

The above is an *operator*. It can be *applied* to an argument, say x:

```
> Delta(sin)(x);
```

$$\sin(x + h) - \sin(x)$$

```
> Delta(f);
```

$$(x \to f(x + h)) - f$$

```
> "(x);
```

$$f(x + h) - f(x)$$

```
> "/h;
```

$$\frac{f(x + h) - f(x)}{h}$$

```
> limit(", h=0);
```

$$D(f)(x)$$

That last result, of course, implicitly assumes that f is differentiable.

3.1.2 Remarks on mathematical operators

Mathematical operators such as the shift operator E_h defined above, the finite divided difference operators, the differentiation operator D, the 'kernel' operators $K : f \rightarrow \int_a^b k(\cdot, x) f(x) \, dx$, and many others provide useful constructs in applied mathematics. By artful (ab)use of notation, useful relations between the operators can be exploited.

$$E_h(f)(x) = f(x + h) = f(x) + D(f)(x)h + \frac{1}{2!}D \circ D(f)(x)h^2 + \cdots$$

$$= (I + hD + \frac{1}{2!}h^2 D^{(2)} + \cdots)(f)(x)$$

$$= \exp(hD)(f)(x)$$

and so $E_h = \exp(hD)$ where we interpret the exponential of the operator hD as the Taylor series of exp with the *powers* replaced by *repeated composition*, so $D^{(2)}$ is interpreted as $D \circ D$, so $D^{(2)}(f)(a) = f''(a)$. Note that the leading 1 in the Taylor series for exp is replaced by the *identity operator* $I : x \rightarrow x$.

Exercises

1. Write a central divided difference operator δ that takes $(x \rightarrow f(x))$ to

$$x \rightarrow \frac{1}{h}(f(x + h/2) - f(x - h/2)) \, .$$

2. Show that the forward divided difference operator Δ that takes $x \rightarrow f(x)$ to $x \rightarrow (f(x+h) - f(x))/h$ is related to the differentiation operator D by $\Delta = (\exp(hD) - 1)/h$. Invert this relationship and use power series in Maple to get a 12th-order accurate finite difference approximation to D.

3. Find a similar relationship between δ and D, invert it, and find an 8th-order finite difference approximation to D using central differences.

4. Using `companion`, `lcoeff`, `Eigenvals`, and `evalf`, write an operator that finds all roots of a given polynomial with numerical coefficients.

3.2 More general procedures

A Maple procedure always returns a value. It is the value of the last statement executed in the procedure before returning. This value may be NULL, which does not print anything on output. The distinction between NULL and 'no value' is academic. One important consequence of a procedure returning NULL is that the environment variables ", "", and """ are not changed. The procedures print and lprint use this deliberately. The procedures solve, fsolve, and dsolve, for example, will return NULL if they find no solution, and sometimes this takes special handling in programs. One simple way to deal with this with solve is to enclose the results from solve in set brackets ({ }), converting a possible NULL value to the empty set.

Examples of simple Maple procedures follow. The first procedure accepts as argument an integer n and returns the value *true* or *false*, depending on whether n is divisible by 17. The second example illustrates the use of for-loops.

Pick a consistent indentation style, such as below. [My editor pointed out that the following procedure breaks the rule I mentioned in the first chapter, of not naming your routines facetiously: the first routine below should be named DivisibleBySeventeen, or something equally informative but shorter: however, these routines don't actually do anything *useful*, and are just intended to exhibit a consistent indentation style. So I feel justified in breaking my rule—which, after all, is not supposed to be 'chipped in stone.']

```
> Fred := proc(n : integer);
>    if n mod 17 = 0 then
>       true
>    else
>       false
>    fi
> end:
> Fred(1);
```

$$false$$

```
> Fred(17);
```

$$true$$

```
> Fred(N);
Error, Fred expects its 1st argument, n, to be of type integer,
but received N
```

```
> Fred(1.5);
Error, Fred expects its 1st argument, n, to be of type integer,
but received 1.5
```

Dominik Gruntz points out that

```
Fred := proc(n:integer)
   evalb(n mod 17 = 0)
end:
```

is a much better program to accomplish the same task. Indeed, if we don't want to do argument-checking, this can be written as the operator n -> evalb(n mod 17=0).

> Ginger Rogers did everything Fred Astaire ever did,
> only backwards and in high heels.
> —Anonymous

```
> Ginger := proc(x : numeric)
>    local s,i,j,k;
>    s := 0;
>    for i from 0 to 5 do
>       for j from -1 to 3 do
>         for k to 3 do
>              s := s + x^(i+j-k)
>          od
>       od
>    od
> end:
> Ginger(0.3);
```

$$349.1084531$$

```
> Ginger(1);
```

$$90$$

```
> Ginger(x);
Error, Ginger expects its 1st argument, x, to be of type numeric,
but received x
```

There is really only one loop construct in Maple. It is a generalized for/while loop, which can have a logical condition as well as a counter. See ?for or ?while for details.

Remark. The punctuation of the above procedures appears to omit some closing semicolons (;) or colons (:). This is deliberate—the last statement in a procedure does not need a semicolon, and likewise neither the last statement in a for-loop nor the last clause in an if-statement needs a terminator. I leave these off when I can, not out of laziness but rather because inserting a statement *after* such a terminating statement can have a larger effect than just the execution of that new statement. The value returned by a for-loop is the value of the last statement executed; likewise, the value returned by a procedure is usually the value of the last statement executed. Adding another statement will change that, and if I didn't mean that to happen I want an error message. This is merely a personal use of this punctuation feature. You may, if you like, always put terminators on your Maple statements—it will make no difference to the program as written. I have been told that this practice goes quite against what is usually taught in CS courses, namely that as much as possible, code should be written so that the insertion of a valid statement either before or after any other valid statement should not generate a syntax error. Please use what works for you.

3.2.1 Structured types

Maple has types, as you have seen from the previous examples. The basic types correspond to the data structures mentioned in Chapter 1. There is a mechanism for querying the type of an expression, the type command. This is especially useful for recognizing complicated 'structured' types. For example, the expression x^2 is of type algebraic, but it is also of type anything^integer.

```
> f := 1 + (x+z)^3 + tan((x+z)^3)*sin(y);
```

$$f := 1 + (x + z)^3 + \tan((x + z)^3)\sin(y)$$

```
> type(f, '+');
```

$$true$$

```
> type(f, algebraic);
```

$$true$$

```
> whattype(f);
```

$$+$$

```
> type(x^2, algebraic);
```

$$true$$

```
> type(x^2, anything);
```

$$true$$

```
> type(x^2, set);
```

$$false$$

```
> type(x^2, anything^posint);
```

$$true$$

Type-checking is one of the simplest and most useful ways of preventing bugs in your program, and of tracking down the ones that occur. If you do not do your own argument type-checking, you run the risk of getting very cryptic error messages on occasion. Prior to Maple V Release 2, type-checking had to be done explicitly by calls to the type routine, but now we can code this concisely as follows.

```
> # The old, Maple V way.
> a := proc(x);
>    if type(x, integer) then
>       2*x
>    else
>       ERROR('expecting integer, got', x)
>    fi
> end:
> # The new, Maple V Release 2 way.
>
> b := proc(x:integer); 2*x end:
>
> a(10);
```

$$20$$

```
> a(Pi);
Error, (in a) expecting integer, got, Pi

> b(10);
```

$$20$$

```
> b(Pi);
Error, b expects its 1st argument, x, to be of type integer, but received Pi
```

Exercises

1. Write a procedure that takes one argument and squares it if it is bigger than 1.

2. Try `subs(17=15, eval(Fred))` and see what you get. This trick is not recommended, because sometimes the results are surprising. Try `subs(1=2, eval(Fred))` and comment.

3. Write an 'Arithmetic-Geometric Mean' operator that takes two numbers a and b and returns $(a + b)/2$ and $(ab)^{1/2}$. [Note: Maple's square root function `sqrt` does some processing before simplifying its input α to $\alpha^{1/2}$. You can save some execution time, not crucial here but perhaps of interest, by directly using $\alpha^{1/2}$.] My operator for this procedure is only 27 characters long, not counting spaces but including the semicolon—is there a better way?

4. Write a procedure `'convert/polyop'` that will convert a polynomial in a variable (e.g., $p = 3 + d + \frac{1}{2}d^2$) to the analogous mathematical operator (e.g., $p = 3I + d + \frac{1}{2}d \circ d$). Be careful of the constant term.

3.3 Data structures

What follows is a very brief overview of Maple's data structures. For a more in-depth look, see (27), and for complete details consult (9). There are many built-in data structures in Maple, many quite different from those of FORTRAN, C, or Pascal. The main data structures are

1. sequences
2. lists
3. sets
4. algebraics
5. unevaluated or inert function calls
6. tables
7. series
8. strings
9. indexed names
10. relations
11. procedure bodies

and the numeric data structures for integers, fractions, floats, and complex numbers. Matrices are currently represented in Maple by arrays, which are a special kind of table.

An 'algebraic' data structure is represented internally as a directed acyclic graph, or DAG for short. For most users, it is best to think of this as simply an expression containing symbols. For example, the expression `1 + (x+z)^3 + tan((x+z)^3)*sin(y)` is of algebraic type, and its DAG is explored in the session below and sketched in Figure 3.1. I remark that the session below is sensitive to *ordering changes*—Maple orders subexpressions in a session-dependent fashion, because Maple uses *address order* for efficiency. This can cause confusion when execution of a Maple script produces different results from previous executions.

```
> f := 1 + (x+z)^3 + tan((x+z)^3)*sin(y);
```

$$f := 1 + (x + z)^3 + \tan\left((x + z)^3\right)\sin(y)$$

```
> nops(f);
```

$$3$$

```
> op(1, f);
```

$$1$$

```
> op(2, f);
```

$$(x + z)^3$$

```
> op(3, f);
```

$$\tan\left((x + z)^3\right)\sin(y)$$

```
> op(1, op(2, f));
```

$$x + z$$

```
> op(1, op(3, f));
```

$$\tan\left((x + z)^3\right)$$

The op command is a useful 'low-level' procedure for picking apart the operands of an object. The command nops counts the number of operands in an object.

Function calls, such as `sin(x)` or `ChebyshevForm(x, A, 12)`, use parentheses. We will see uses for unevaluated function calls as data structures later.

Tables use parentheses in their creation, as in `T := table()` or `A := array(0..3)`, and square brackets to reference individual entries, as in `A[3]`.

Figure 3.1 The directed acyclic graph of $1 + (x + z)^3 + \tan((x + z)^3)\sin(y)$

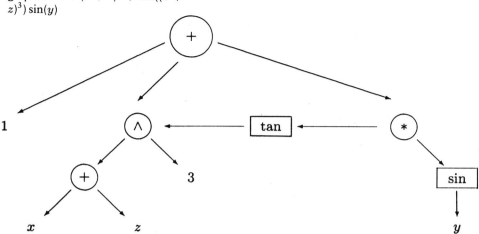

Sequences consist of two or more objects separated by commas.

```
> e_seq := 1,2,3,4:
```

Lists are simply expression sequences in square brackets. The i-th element of an expression sequence or list can be selected as though it were in an array (though internally this is done by linear search instead of address calculation as is done for tables, and hence is less efficient).

```
> L := [1,2,3,4]:
> M := [2,3,4,1]:
```

L and M in the above example are different, since order is important for lists. To select the third entry in each, use e_seq[3], L[3], and M[3].

Sets, on the other hand, use curly brackets, and order is unimportant:

```
> S := {1, 2, 3, 4}:
> T := {2, 3, 4, 1}:
```

In this example, S and T define the *same set* and indeed Maple will detect this (by hashing) and only store one copy of this set. [This can be verified by using the addressof function to locate the data structures in memory.]

The series data structure arises only from a call to the series, taylor, or asympt commands. Manipulate these by further calls to those commands, or by converting them to sums of terms. The series data structure is a *sparse* data structure: terms with zero coefficients are not stored. For details, see ?type[series].

```
> series( sin(x), x );
```

$$x - \frac{1}{6}x^3 + \frac{1}{120}x^5 + O(x^6)$$

```
> series( cos(x), x );
```

$$1 - \frac{1}{2} x^2 + \frac{1}{24} x^4 + O(x^6)$$

```
> " + "";
```

$$\left(1 - \frac{1}{2} x^2 + \frac{1}{24} x^4 + O(x^6)\right) + \left(x - \frac{1}{6} x^3 + \frac{1}{120} x^5 + O(x^6)\right)$$

```
> whattype(");
```

$$+$$

```
> series("", x);
```

$$1 + x - \frac{1}{2} x^2 - \frac{1}{6} x^3 + \frac{1}{24} x^4 + \frac{1}{120} x^5 + O(x^6)$$

```
> whattype(");
```

$$\text{series}$$

```
> convert("", polynom);
```

$$1 + x - \frac{1}{2} x^2 - \frac{1}{6} x^3 + \frac{1}{24} x^4 + \frac{1}{120} x^5$$

Sometimes the result from `series` is not of type `series`. This occurs particularly when the result is a Puiseux series.

```
> series( sin(x^(1/2)), x);
```

$$\sqrt{x} - \frac{1}{6} x^{3/2} + \frac{1}{120} x^{5/2} - \frac{1}{5040} x^{7/2} + \frac{1}{362880} x^{9/2} - \frac{1}{39916800} x^{11/2} + O(x^6)$$

```
> whattype(");
```

$$+$$

In this case the usual method of converting a series to an algebraic object does not work.

```
> convert("", polynom);
```

$$\sqrt{x} - \frac{1}{6}\,x^{3/2} + \frac{1}{120}\,x^{5/2} - \frac{1}{5040}\,x^{7/2} + \frac{1}{362880}\,x^{9/2} - \frac{1}{39916800}\,x^{11/2} + O(x^6)$$

Instead, one has to resort to a trick. We substitute 0 (zero) for O (oh).

```
> subs(O=0, ");
```

$$\sqrt{x} - \frac{1}{6}\,x^{3/2} + \frac{1}{120}\,x^{5/2} - \frac{1}{5040}\,x^{7/2} + \frac{1}{362880}\,x^{9/2} - \frac{1}{39916800}\,x^{11/2} + 0(x^6)$$

We must evaluate after a substitution, usually by referring to the object again.

```
> ";
```

$$\sqrt{x} - \frac{1}{6}\,x^{3/2} + \frac{1}{120}\,x^{5/2} - \frac{1}{5040}\,x^{7/2} + \frac{1}{362880}\,x^{9/2} - \frac{1}{39916800}\,x^{11/2}$$

That trick works in the following way. Since numbers are also operators (they are constant operators: for example, $1(x) \rightarrow 1$ for any x), we can remove the O symbol by substituting zero for it. The resulting object is now an ordinary Maple algebraic. One can see from the first part of this example that series data structures cannot be manipulated without repeated calls to `series`.

Strings are reasonably important in Maple. They are used mainly as names of things and as messages. A Maple string can be used as a variable name; most often these are simple, unquoted strings. But if your variable name contains /, as in `'diff/T'`, then you must enclose your variable name in string quotes. Use string quotes for strings containing filenames, because without them the period (.), which is Maple's concatenation operator, will turn `file.ext` into `fileext`, which isn't what you want.

Indexed names are objects that look like `Database[1, 2, drawer]` or `A[3]`. You need not explicitly create a table before you start using indices. Maple will automatically create a table if it has to. Indexed names are used most often for matrices and vectors. To ensure that Maple creates the right sort of table, an explicit creation before assignment is recommended.

```
> x := T[3];
```

$$x := T[3]$$

```
> eval(T);
```

$$T$$

```
> T[5] := y;
```

$$T[5] := y$$

```
> eval(T);
```

$$table([5 = y])$$

Maple did not create a table for T until we assigned something to T[5].

```
> A := linalg[matrix](3, 3, (i,j)->i^(j-1));
```

$$A := \begin{bmatrix} 1 & 1 & 1 \\ 1 & 2 & 4 \\ 1 & 3 & 9 \end{bmatrix}$$

```
> A[1,2];
```

$$1$$

```
> B := linalg[vector](2);
```

$$B := \text{array}(\,1..2, [\;]\,)$$

```
> print(B);
```

$$[B_1, \quad B_2]$$

Creation of a matrix lets us access its entries by indexing. There is an interesting trick here: linalg[matrix] is itself using tables. The table is linalg, and each entry is a procedure body: in the above example, the index was matrix, which is the name associated to the procedure stored in the table entry linalg[matrix]. Well, almost; actually the entry is the readlib-definition of the procedure.

```
> C := linalg[matrix](2,2);
```

$$C := \text{array}(\,1..2, 1..2, [\;]\,)$$

```
> linalg[linsolve](C, B);
```

$$\left[-\frac{C_{1,2}\,B_2 - B_1\,C_{2,2}}{C_{1,1}\,C_{2,2} - C_{2,1}\,C_{1,2}},\; \frac{C_{1,1}\,B_2 - C_{2,1}\,B_1}{C_{1,1}\,C_{2,2} - C_{2,1}\,C_{1,2}} \right]$$

Notice that the answer contains indexed names. Type name includes strings and indexed names.

The numeric data structures (integer, fraction, rational, float, hardware float, and complex numeric) are largely transparent in their uses. Floats are usually contagious, which means that if one element of a numeric structure is a float, then automatically the whole structure is converted to floating point, if possible.

Hardware floats are automatically converted to Maple floats on return from evalhf. This vitiates most efficiency gains that a user could hope for in using evalhf, so until facilities for direct manipulation (or at least for unconverted storage) are provided, evalhf is best left for internal Maple use.

3.4 Local vs. global vs. environment variables

Local variables are local to a procedure, are unavailable to subroutines called by that procedure, and disappear once that procedure completes its execution (unless they are 'exported'). Using local variables is an excellent way of hiding information irrelevant to other procedures, and freeing up short names for use elsewhere. In the procedure Ginger (p. 131), the variables i,j,k, and s are all local, and that example shows how to declare local variables in procedures. To declare local variables in an operator, you must use angle-bracket notation: for example, the shift operator that we have already seen required a local variable x.

```
Eh := <unapply(f(x+h), x) | f | x >:
```

The first | indicates the argument(s) to the operator, and the second defines the local variable(s) if any (here, one local variable x is declared). As another example, the operator

```
<unapply(convert([seq(x^k, k=0..n)], '+'), x) | n | x,k >
```

has two local variables, x and k, and one argument to the operator, n.

3.4.1 Exporting local variables

Sometimes it is useful to export local variables, which are unique (although two or more different ones may look the same) and difficult to 'touch.' An example of an application of exporting local variables is detailed in (12, vol. 2) in the Chebyshev Forms package. Another is to use exported local variable names to uniquely identify 'infinities' so Maple's automatic simplification does not erroneously cancel them, according to an idea of Dominik Gruntz.

An amusing demonstration of exporting local names was given on USENET in 1993 by Frederic W. Chapman, along the following lines. [This example presumes some knowledge of *Star Trek: The Next Generation*, but hopefully the point will still get across even if the reader is not a Trekker.]

```
> Holodeck := proc() local Moriarty; Moriarty end:
> Holodeck();
```
$$Moriarty$$

Moriarty has escaped from the Holodeck!

```
> Moriarty - ";
```
$$Moriarty - Moriarty$$

He is *different* from any Moriarty that the user can type in. But nonetheless he can be touched:

```
> evalb(""=Moriarty);                          # Should be false
```
$$false$$

```
> ``.""""-``.Moriarty;  # Should be zero, as we are subtracting strings
```
$$0$$

That last trick is attributed to Dominik Gruntz. We concatenate the NULL string to the local variable name, and the result (a string) is independent of where the variable came from: local or global, it doesn't matter. Different instances of Moriarty are different (though they look the same):

```
> Holodeck();
```
$$Moriarty$$

```
> Holodeck();
```
$$Moriarty$$

```
> ""-";  # If they were the same this would automatically be zero
```
$$Moriarty - Moriarty$$

```
> evalb(``.""""=``.."");
```
$$true$$

3.4.2 Global variables

Global variables are available to anything that references their name. If a local variable has the same name as a global variable, the local variable 'masks' the global one and that global variable is unavailable in that procedure. Use of global variables should be restricted as much as possible, since they can conflict with other user's names for their variables (or indeed your own). If you must use a global variable, make its name longer than necessary to help avoid conflicts. *Starting in Maple V Release 3, global variables must be declared global in procedures.* Otherwise they may be automatically declared as local, if they occur on the left-hand side of an assignment or in a loop, and this will mean that some procedures will work differently than they did in earlier versions. Usually if this is the case, the procedure in question was less robust than it should have been.

Most system global variables start with the character _, e.g., _NCRule. Do not use variable names starting with this character. [Except for 'environment' variables—see below.]

It is often difficult to make sure you have declared all your variables as local, and unexpected globals can be a time bomb for your routines—they may work fine for months, and suddenly fail if you assign a global variable with the same name as the inadvertent global in your routine. This is the reason for the new automatic declaration of global variables in Maple V Release 3. The program utility mint, called outside of Maple, helps to deal with this. It will check to see if there are any global variables in your procedure, and check to see that all locals are used. It checks for other Maple syntax errors as well, which can be much quicker than running Maple to find the syntax errors. So, *use* mint *to check your programs.*

3.4.3 Environment variables

Environment variables are, roughly speaking, global variables that are automatically reset on exit from a procedure. There are several built-in environment variables in Maple and facilities for adding your own. The built-in environment variables are Digits, Normalizer, Testzero, mod, printlevel, and the ditto commands ", "", and """. Finally, any variable beginning with _Env is an environment variable, so you can define your own. See ?environment for more details.

```
> fu := proc(x);
>    bah(x);
>    _Envmyjunk := 3;
>    bah(x);
> end:
```

```
> bah:= proc(y);
>    if _Envmyjunk = 5 then
>       print('It's 5, I tell you!')
>    elif _Envmyjunk = 3 then
>       print('No, it's 3!')
>    else
>       print('I don't know what it is, boss.')
>    fi
> end:

> _Envmyjunk := 5;
```

$$\_Envmyjunk := 5$$

```
> fu(throgmorton);
```

$$It's\ 5,\ I\ tell\ you!$$

$$No,\ it's\ 3!$$

In the above, we see that _Envmyjunk is reset *automatically* on exit from fu.

```
> _Envmyjunk;
```

$$5$$

The series command now makes use of the Testzero environment variable, which means that you can select the normalizer to use on computation of series coefficients. This fixes several outstanding bugs in the computation of series, that arose because the leading coefficient of a series was not previously recognized as infinity. The following example illustrates the meaning and use of this variable. It also gives our first illustration of the difficulties associated with 'option remember,' which will be discussed more fully in the next section.

The variable Testzero is initially set to something that uses normal, but should probably use Normalizer instead. This may already be fixed by the time this book is published.

The *residue* of a function $f(z)$ at a point $z = a$ is the coefficient of $1/(z - a)$ in the Laurent series expansion of f at $z = a$. It is used in the computation of contour integrals, among other things.

```
> Testzero := x -> evalb(Normalizer(x)=0);
```

$$Testzero := x \rightarrow evalb(\,Normalizer(\,x\,) = 0\,)$$

```
> p := x^3 + x + 1;
```

$$p := x^3 + x + 1$$

```
> alias(alpha=RootOf(p, x));
```

$$I, \alpha$$

See (12, vol. 1) for more details on `RootOf`.

```
> readlib(residue):
> residue( 1/p, x=alpha);
```

$$0$$

This is incorrect, since α is a root of p. But `residue` calls `series`, which uses `normal`, which does not recognize that $p(\alpha) = 0$ (because `normal` is not strong enough).

```
> series(1/p, x=alpha, 3);
```

$$\frac{1}{\alpha + 1 + \alpha^3} - \frac{1 + 3\,\alpha^2}{(\alpha + 1 + \alpha^3)^2}\,(x - \alpha) + \frac{-3\,\dfrac{\alpha}{\alpha + 1 + \alpha^3} + \dfrac{\left(1 + 3\,\alpha^2\right)^2}{(\alpha + 1 + \alpha^3)^2}}{\alpha + 1 + \alpha^3}\,(x - \alpha)^2$$
$$+ O\left((x - \alpha)^3\right)$$

That answer is incorrect.

```
> Normalizer := x -> normal(simplify(x));
```

$$\text{Normalizer} := x \rightarrow \text{normal}(\,\text{simplify}(\,x\,)\,)$$

```
> readlib(forget):
> forget(series);
```

That last command wipes out the remember table for `series`.

```
> series(1/p, x=alpha, 3);
```

$$\frac{1}{1 + 3\,\alpha^2}\,(x - \alpha)^{-1} + 3\,\frac{\left(-\dfrac{4}{31} - \dfrac{6}{31}\,\alpha^2 + \dfrac{9}{31}\,\alpha\right)\alpha}{1 + 3\,\alpha^2} + O(x - \alpha)$$

```
> residue(1/p, x=alpha);
```

$$0$$

Maple (in the routine `residue`) remembered the incorrect result from before.

```
> forget(residue);
> residue(1/p, x=alpha);
```

$$\frac{1}{1 + 3\,\alpha^2}$$

That's better.

```
> simplify(");
```

$$\frac{4}{31} + \frac{6}{31}\,\alpha^2 - \frac{9}{31}\,\alpha$$

Exercises

1. Compute the residues of $q = (1 + z)/(1 + z + z^2 + z^3)$.
2. Compute the series expansion of q about $z_0 = \exp(i\pi/4)$.

3.5 Recursion and 'option remember'

Recursion allows the programmer to be lazy. To be fair, it does allow
compact programs, and programmer time is important, too. However, it
can lead to extremely wasteful usage of computer resources (to the point
where the desired task cannot be performed). The classical example of this
is the recursive calculation of the Fibonnaci numbers.

```
> fib_stupid := proc(n:integer);
>    if n<=1 then
>       1
>    else
>       fib_stupid(n-1) + fib_stupid(n-2)
>    fi
> end:

> fib_ok := proc(n:integer)
>    option remember;
```

```
>    if n<=1 then
>      1
>    else
>        fib_ok(n-1) + fib_ok(n-2)
>    fi
> end:
```

Note that there is no semicolon between the `proc(args)` and the `option remember`.

```
> st := time():
> fib_stupid(15);
```

$$987$$

```
> time_1 := time() - st;
```

$$time_1 := 4.000$$

The time for `fib_stupid` is essentially proportional to the value of `fib_stupid`, which grows exponentially with n. The Maple `time()` function returns the time in seconds from some reference time.

```
> st := time();
```

$$st := 5.000$$

```
> fib_ok(15);
```

$$987$$

```
> time_2 := time() - st;
```

$$time_2 := 0$$

So small it doesn't register here. This computing time grows linearly.

```
> fib_ok(100);
```

$$573147844013817084101$$

`fib_stupid(100);` cannot be computed on this machine (a 25Mhz IBM PC clone)—it would take more than a billion years, assuming enough memory could be found for it. The Maple construct `option remember` can correct this exponential growth in time and space requirements, reducing it to linear growth. This is a good thing, occasionally, and a bad thing if overused, because memory requirements are often much higher in a procedure that uses `option remember` than a correctly written procedure that doesn't. Compare `fib_ok` to the built-in `combinat[fibonacci]` procedure, which is *much* more efficient, and does *not* use `option remember`.

Warning: option remember *can cause bugs.* option remember does *not* take into account the values of global or environment variables (except in evalf which takes special care with Digits). So if your proc uses global or environment variables, *don't* use option remember. This is why we had to use forget for series when we changed the environment variable Normalizer earlier. *This is not foolproof—you may have to call* forget *on every subroutine that has* option remember, *or else restart the Maple session using* restart, *to make sure that everything has really been forgotten.* The routine forget has been improved for Maple V Release 3.

Finally, option remember and tables/arrays/vectors do not mix. Assignment of table entries does not change the name of the table, hence option remember won't notice if any table entries have been changed.

```
> f := proc(t:array(anything))
>    option remember;
>    if t[1] = 5 then
>       1
>    else
>       0
>    fi
> end:

> t := array(1..3);
```

$$t := array(1..3, [])$$

```
> t[1] := 5;
```

$$t[1] := 5$$

```
> f(t);
```

$$1$$

```
> t[1] := 'not 5, that's for sure';
```

$$t[1] := not\ 5,\ that's\ for\ sure$$

```
> f(t);
```

$$1$$

f has remembered the previous result, even though it is not what we wanted. *Don't mix tables/arrays/vectors with* option remember.

Final advice on option remember: Use it if you want (it can save you time), but be careful. If you can spare the effort, see if you can write your procedures without it. Also, be aware of option system, which modifies option remember so that that garbage collection clears the remember table frequently.

3.6   Variable number or type of arguments

The number of actual arguments to Maple procs can vary at run time. Some functions must have a minimum number of arguments—for example,

```
> f := proc(x, y);
>    if x < 1 then
>       3
>    elif y < 1 then
>       5
>    else
>       0
>    fi
> end:
```

If this routine is called with no arguments, by f(), then an error message will ensue (too few arguments). If it is called with only one argument, one of two things can happen, depending on the value of the argument: if f(1/2) is the call, then the result will be 3; whereas if f(2) is the call, the first if statement will fail and the second will execute: at that point Maple will discover that it has no value for y and complain (again, too few arguments).

For times when you want to deal with the case when there are *more* arguments than specified in the procedure header, you can use args and nargs. At run time, the variable nargs contains the number of actual arguments to the procedure (which may vary from invocation to invocation) and the expression sequence args contains the actual arguments: The i-th argument can be referenced as args[i]. See also the next section for more examples of the use of nargs and args.

Structured types will allow type-checking for more than one type of input argument. The following procedure will accept arguments of type name or an equation where the left-hand side is a name and the right-hand side is a range with constant endpoints.

This routine supplies a default range of integration.

```
> my_int := proc(f, x:{name, name=constant..constant});
>    if type(x, name) then
>       int(f, x=0..1)
>    else
>       int(f, x)
>    fi
> end:
```

```
> my_int( sin(Pi*x), x);
```

$$2\,\frac{1}{\pi}$$

```
> my_int( cos(Pi*x/2), x=-1..1);
```

$$4\,\frac{1}{\pi}$$

```
> my_int( sin(Pi*x), x=0..4);
```

$$0$$

The following extended example shows a nontrivial procedure that uses type-checking and a variable number of arguments. It converts a given column vector, or a column-row vector pair, into a Hankel matrix, which is a matrix constant along anti-diagonals. This routine is sufficiently useful that a help file is also provided.

```
Hankel := proc(C:vector, R:vector) local i,j,n,h;
  n := linalg[vectdim](C);
  h := linalg[matrix](n, n, 0);

  if n < 2 then ERROR('Don't be silly.'); fi;

  for j to n do
    for i to n-j+1 do
      h[i,j] := C[i+j-1];
    od;
  od;

  if nargs > 1 and linalg[vectdim](R)=n then

    if C[n] <> R[1] then
      print('Column wins anti-diagonal conflict', R[1]);
    fi;

    for j from 2 to n do
      for i from n by -1 to n-j+2 do
        h[i,j] := R[j+i-n];
      od;
    od;

  fi;
  eval(h)
end:
```

3.7   Creating your own 'help file'

The following is an abridged version of the help file for `Hankel`. The complete text of the help file follows after, in a slightly more human-readable format, with all the left quotes and commas stripped out (exactly as the Maple help utility would display it).

```
'help/text/Hankel' := TEXT(
'FUNCTION:  Hankel    ---   transform an input vector or vectors',
'                           to a Hankel matrix, which is a matrix',
'                           constant along anti-diagonals.',
'CALLING SEQUENCE:',
'   ',
'  Hankel(C);      or',
'  Hankel(C, R);',
'   ',
```

many lines omitted

```
'   '
'SEE ALSO: linalg[toeplitz]',
'   '
):
```

A help file in Maple is a variable named `'help/text/...'`, which must contain a TEXT object. A TEXT object is merely an unevaluated function call, whose arguments are Maple strings. When the help file is displayed, each argument will be printed on a separate line.

To create your own help file, you must create a similar object. The routine `makehelp` will convert an ordinary text file to a help file. See `?makehelp` for details.

The complete help file for `Hankel` follows.

```
FUNCTION:  Hankel    ---   transform an input vector or vectors
                           to a Hankel matrix, which is a matrix
                           constant along anti-diagonals.

CALLING SEQUENCE:

  Hankel(C);      or
  Hankel(C, R);

  where C is the vector of elements intended for the first
  column of the Hankel matrix, and R is a vector intended for
  the last row of the matrix.  If R is not specified, the subdiagonals
  will all be set to zero.

REFERENCE:

  This routine was adapted from the MATLAB routine of the same name
  written by John Little.

  For the theory of Hankel matrices, see 'Matrix Computations' by
  Golub and Van Loan.
```

EXAMPLES:

```
> C := linalg[vector]([1,2,3,4]);
```

$$C := [1, \ 2, \ 3, \ 4]$$

```
> Hankel(C);
```

$$\begin{bmatrix} 1 & 2 & 3 & 4 \\ 2 & 3 & 4 & 0 \\ 3 & 4 & 0 & 0 \\ 4 & 0 & 0 & 0 \end{bmatrix}$$

```
> R := linalg[vector]([4,5,6,7]);
```

$$R := [4, \ 5, \ 6, \ 7]$$

```
> Hankel(C, R);
```

$$\begin{bmatrix} 1 & 2 & 3 & 4 \\ 2 & 3 & 4 & 5 \\ 3 & 4 & 5 & 6 \\ 4 & 5 & 6 & 7 \end{bmatrix}$$

```
> Col := n -> linalg[vector]([seq(a^k, k=0..n-1)]);
```

$$\text{Col} := n \rightarrow \text{local } k; \ \text{linalg}_{\text{vector}}([\,\text{seq}(\,a^k, k = 0..n-1\,)\,])$$

```
> Row := n -> linalg[vector]([seq(a^(n-k), k=1..n)]);
```

$$\text{Row} := n \rightarrow \text{local } k; \ \text{linalg}_{\text{vector}}([\,\text{seq}(\,a^{(n-k)}, k = 1..n\,)\,])$$

```
> C := Col(3);
```

$$C := [1, \ a, \ a^2]$$

```
> R := Row(3);
```

$$R := [\,a^2,\ a,\ 1\,]$$

```
> Hankel(C, R);
```

$$\begin{bmatrix} 1 & a & a^2 \\ a & a^2 & a \\ a^2 & a & 1 \end{bmatrix}$$

```
> C := Col(5):
> R := Row(5):
> R[1] := 'Bzzlplplplplt!';
```

$$R_1 := \text{Bzzlplplplplt!}$$

```
> Hankel(C, R);
```

Column wins anti-diagonal conflict, Bzzlplplplplt!

$$\begin{bmatrix} 1 & a & a^2 & a^3 & a^4 \\ a & a^2 & a^3 & a^4 & a^3 \\ a^2 & a^3 & a^4 & a^3 & a^2 \\ a^3 & a^4 & a^3 & a^2 & a \\ a^4 & a^3 & a^2 & a & 1 \end{bmatrix}$$

```
SEE ALSO: linalg[toeplitz]
```

If you create lots of Maple programs, you should create help files for each of them. This can now be done easily with the makehelp procedure. See ?makehelp for details.

Another thing that is useful if you are creating many programs is the Maple Archiver or march. This helps you to manage archives of .m files. See ?march for details.

3.8 Returning more than one result

Maple is set up to return only one result from a procedure. Mind you, that result can be an array, table, list, set, or expression sequence, so the consequences of this restriction are not severe. However, sometimes you want to pass a name to a procedure and have it assigned in the procedure (as is common in FORTRAN). This is possible, and the following example shows how. Pay particular attention to the use here of right or forward quotes, which prevent evaluation of the names to be assigned.

The procedure mynormal below returns an expression sequence as an answer, the elements of which are supposed to be the numerator and denominator of a quotient. If asked, this routine also returns the gcd of the two quantities.

```
> mynormal := proc(p, q, x) local g,pc,qc;
>     g := gcd(p, q);
>     if nargs > 3 and type(args[4], name) then
>       assign(args[4], g);
>     fi:
>     rem(p, g, x, pc);
>     rem(q, g, x, qc);
>     pc,qc
> end:
```

This is a 'recursive' example, since the subroutine rem (and its complementary routine quo) use the trick we are trying to explain here. See ?rem for details.

```
> P := expand( (x+1)^3*(x+2)^2*(x+3) );
```

$$P := x^6 + 10\,x^5 + 40\,x^4 + 82\,x^3 + 91\,x^2 + 52\,x + 12$$

```
> Q := diff(P, x);
```

$$Q := 6\,x^5 + 50\,x^4 + 160\,x^3 + 246\,x^2 + 182\,x + 52$$

```
> mynormal(P, Q, x);
```

$$x^3 + 6\,x^2 + 11\,x + 6, 6\,x^2 + 26\,x + 26$$

```
> mynormal(P, Q, x, G);
```

$$x^3 + 6\,x^2 + 11\,x + 6, 6\,x^2 + 26\,x + 26$$

This worked because G was unassigned.

```
> G;
```

$$x^3 + 4\,x^2 + 5\,x + 2$$

```
> mynormal(P, Q, x, 3);
```

$$x^3 + 6\,x^2 + 11\,x + 6, 6\,x^2 + 26\,x + 26$$

This time, even though there were four arguments present, the fourth argument was not of type name and so no assignment was attempted. If the assignment had been attempted, we would have received an error message. Perhaps it is desirable that an error message be generated because G might inherit a previous value, erroneous in the present context. So the if statement can be changed, to give the following routine:

```
> mynormal2 := proc(p, q, x) local g,pc,qc;
>    g := gcd(p, q);
>    if nargs > 3 and type(args[4], name) then
>      assign(args[4], g);
>    elif nargs > 3 then
>      ERROR('I am very sorry, but I cannot assign a value to
>      ', args[4])
>    fi:
>    rem(p, g, x, pc);
>    rem(q, g, x, qc);
>    pc,qc
> end:

> mynormal2(P, Q, x, G);
Error, (in mynormal2) I am very sorry, but I cannot assign a value
to , x^3+4*x^2+5*x+2
```

Note that if we do not check if args[4] is of type name, Maple will give a less understandable error message, such as "Illegal use of an object as name," if we try to assign a value to $x^3 + 4x^2 + 5x + 2$.

A useful strategy to deal with code like the original mynormal (some library routines use this method of passing values: e.g., rem and quo) is to pass unevaluated names. That is, use quotes whenever there is a chance that the names may have values. Notice that in the calls to rem in mynormal and mynormal2 quotes around pc and qc are not used—this is because they are unnecessary as the local variables obviously have no other values, and on each invocation get created anew.

Even so, if this code were later to be modified, it is possible that values could be assigned to these local variables—in that case, an error would occur.

```
> mynormal(P, (x+1)*Q, x, 'G');
```

$$x^2 + 5x + 6, 6x^2 + 26x + 26$$

This 'unevaluation' says to mynormal to use the *name* G and not its value.

```
> G;
```

$$x^4 + 5x^3 + 9x^2 + 7x + 2$$

Of course this works for mynormal2, also.

```
> mynormal2(P, (x+3)*Q, x, 'G');
```

$$x^2 + 3x + 2, 6x^2 + 26x + 26$$

```
> G;
```

$$x^4 + 7x^3 + 17x^2 + 17x + 6$$

3.9 Debugging Maple programs

The trace facility allows you to trace the execution of your procedures (or, indeed, any library procedure). The printlevel variable, when set to values higher than 1, allows a more detailed view of the execution of a program.

```
> mork := proc(a:integer) local k;
>    k := a^2;
>    mindy(k + a)
> end:

> mindy := proc( b: integer) local m;
>    m := b/2;
>    if m > 10 then
>       1
>    else
>       1 + mork(b)
>    fi
> end:

> mork(1);
```

3

```
> trace(mork);
```

$$mork$$

```
> mork(1);
{--> enter mork, args = 1
```

$$k := 1$$

```
{--> enter mork, args = 2
```

$$k := 4$$

```
{--> enter mork, args = 6
```

$$k := 36$$

$$1$$

```
<-- exit mork (now in mindy) = 1}
```

$$2$$

```
<-- exit mork (now in mindy) = 2}
```

$$3$$

```
<-- exit mork (now at top level) = 3}
```

$$3$$

```
> untrace(mork);
```

$$mork$$

```
> trace(mindy);
```

$$mindy$$

```
> mork(1);
{--> enter mindy, args = 2
```

$$m := 1$$

```
{--> enter mindy, args = 6
```

$$m := 3$$

```
{--> enter mindy, args = 42
```

$$m := 21$$

$$1$$

```
<-- exit mindy (now in mork) = 1}
```

$$2$$

```
<-- exit mindy (now in mork) = 2}
```

$$3$$

```
<-- exit mindy (now in mork) = 3}
```

$$3$$

```
> untrace(mindy);
```

$$mindy$$

```
> printlevel := 1000;
```

$$printlevel := 1000$$

```
> mork(1);
{--> enter mork, args = 1
```

$$k := 1$$

```
{--> enter mindy, args = 2
```

$$m := 1$$

```
{--> enter mork, args = 2
```

$$k := 4$$

```
{--> enter mindy, args = 6
```

$$m := 3$$

```
{--> enter mork, args = 6
```

$$k := 36$$

```
{--> enter mindy, args = 42
```

$$m := 21$$

$$1$$

```
<-- exit mindy (now in mork) = 1}
```

$$1$$

```
<-- exit mork (now in mindy) = 1}
```

$$2$$

```
<-- exit mindy (now in mork) = 2}
```

$$2$$

```
<-- exit mork (now in mindy) = 2}
```

$$3$$

```
<-- exit mindy (now in mork) = 3}
```

$$3$$

```
<-- exit mork (now at top level) = 3}
```

$$3$$

If you want to provide the user with extra information about the progress of the solution, instead of debug a procedure, you can use the `infolevel` and `userinfo` facilities. These are extensively used in the Maple library so you can examine the progress (and the methods used) of several library routines, such as `int` and `dsolve`.

Exercises

1. Write a procedure that prints out (in a nice format) the amount of CPU time Maple has used so far, together with a report of how much memory it has used. See `?status`.

2. Write a procedure that accepts as input a matrix A with numeric entries, figures out the dimension of the matrix, and then uses the `linalg` routine `GramSchmidt` to compute an orthogonal matrix Q and an upper-triangular matrix R such that $A = QR$. The Gram-Schmidt procedure is known to be unstable numerically; look up 'modified Gram-Schmidt' and implement that instead.

3. Read all of (34). This will give you a more in-depth view of what you have just read, and more details on the computer-science aspects of Maple programming.

4. Do all the exercises in (34).

5. Read (41).

6. Compare the results of `mork(1);` and `mork(1):` when `printlevel := 100`. This difference is useful for quickly localizing problems. This exercise was suggested by Dave Hare.

3.10 Sample Maple programs

What follows are a few final sample Maple programs that you may use as templates for your own programs and to give you ideas. They are not intended to be examples of 'programming gems,' and I offer them only as working programs that I wrote for actual use.

3.10.1 Parametric solution of algebraic equations

The following procedure uses the trick of substituting $y = tx$ into an algebraic equation $f(x, y) = 0$ to get a parametric solution of the equation. This is useful for plotting purposes or for integration, differentiation, and series. The program is as follows.

```
# FUNCTION:  parsolve  ---  Solve implicit equations parametrically
# RMC February 1994
#
# Basic idea:  Use the trick  y = tx to solve f(x,y) = 0.
# MAINTENANCE HISTORY:
#
# Original version February 1994.
# Fixed x, y bugs June 1994.
# CALLING SEQUENCE:  parsolve(f(x,y), [x,y], t);
# INPUT:  f(x,y)  --- an expression in two variables.
#         [x,y]   --- list (or set) of the two variables.
#         t       --- desired name of the independent variable.
# OUTPUT: set of solutions, labelled with the variable names.
# SIDE EFFECTS:  none.
# GLOBAL VARIABLES:  none.
# KNOWN BUGS/WEAKNESSES: Technique is heuristic and often fails.
# REFERENCES: G. H. Hardy, Pure Mathematics, Cambridge, 1952.
#
# Miscellaneous remarks:  Remarkably powerful heuristic.  It
# succeeds for plotting even in the case of some singular points;
# in that sense it is complementary to some algorithmic techniques
# for finding rational parameterizations of genus 0 curves.
#
parsolve := proc(f, xy:{ list(name), set(name) }, t:name)
   local p,x,y;

   x := xy[1];
   y := xy[2];

   # Throw away the zero solution, if it is there, because it
   # is not interesting: we are looking for curves, not points.

   p := {solve(subs(y=t*x, f), x)} minus {0};
```

```
    map( (xi, u, xx, yy)->{xx=xi, yy=u*xi}, p, t, x, y)

  end:
```

The input to this procedure is the algebraic equation to be solved, f, represented as an expression in two variables, which we represent as x and y in the procedure but may have any distinct names for the actual arguments. We also input the actual names of the variables as a list or a set of names, and then finally the name of the parameter (represented as t in the procedure).

The first statements select the variable names from the input list xy.

We then solve the equation with y replaced by tx for x. The call to solve is wrapped in set brackets, so multiple solutions returned by solve will be pared down to only one solution, and so that the NULL solution will be transformed to the empty set. We subtract the zero solution from this solution set because the solution $x = 0$, $y = 0$, if it occurs, is always uninteresting in this context (because it describes a point, not a curve).

The call to map takes each solution for x found by solve and prepares the solution pair $\{x, y\}$ from that, using $y = tx$. We use the multiple-argument form of map so we can pass the actual name of the parameters to it; if we just had instead

```
    map( xi->{xi,t*xi}, p);
```

then the t in that would be *global*, and not the t that we had intended. Similarly for the names of the x and y variables. This explicit passing of parameters is necessary because Maple does not have 'nested lexical scopes.'

Here are some examples of this procedure's use.

```
  > parsolve( u^2 + v^2 = a^2, [u,v], t);
```

$$\left\{\left\{u = \frac{a}{\sqrt{1+t^2}}, v = \frac{ta}{\sqrt{1+t^2}}\right\}, \left\{v = -\frac{ta}{\sqrt{1+t^2}}, u = -\frac{a}{\sqrt{1+t^2}}\right\}\right\}$$

So this procedure allows us to find a pair of parametric representations of a circle of radius a. Knowing that trigonometric functions give us a better representation, we could, if we desired, set $\cos\theta = 1/\sqrt{1+t^2}$ to get a better parameterization from this one.

We can also do circles centered elsewhere:

```
  > parsolve( s^2 + s*t + t^2 = a^2, [s,t], u);
```

$$\left\{\left\{t = \frac{ua}{\sqrt{1+u+u^2}}, s = \frac{a}{\sqrt{1+u+u^2}}\right\},\right.$$
$$\left.\left\{s = -\frac{a}{\sqrt{1+u+u^2}}, t = -\frac{ua}{\sqrt{1+u+u^2}}\right\}\right\}$$

And now the Folium of Descartes:

```
> parsolve( x^3 - 3*a*x*y + y^3 = 0, [x,y], t);
```

$$\left\{\left\{x = 3\,\frac{a\,t}{1+t^3}, y = 3\,\frac{t^2\,a}{1+t^3}\right\}\right\}$$

```
> assign("[1]);
```

If we wish to plot it, we must nondimensionalize. The plot below is a bit rough, and is not printed here. It is left to the exercises to explain why the plot is so rough.

```
> plot([x/a, y/a, t=-infinity..infinity], view=[-2..2, -2..2]);
> x := 'x'; y := 'y';
```

$$x := x$$

$$y := y$$

Now a generalization of that Folium.

```
> parsolve( x^5 - 5*x*y^3 + y^5, [x,y], u);
```

$$\left\{\left\{y = 5\,\frac{u^4}{1+u^5}, x = 5\,\frac{u^3}{1+u^5}\right\}\right\}$$

Now we show some limitations of this approach. We attempt to solve a random degree 5 polynomial in x and y.

```
> f := randpoly([x,y], degree=5, sparse);
```

$$f := 79\,y^5 + 56\,x\,y + 49\,y + 63\,x^3 + 57\,x^3\,y^2 - 59\,x^2\,y^3$$

```
> parsolve(f, [x,y], t);
```

$$\{\{y = t\,\mathrm{RootOf}((\,79\,t^5 + 57\,t^2 - 59\,t^3\,)\,\_Z^4 + 63\,\_Z^2 + 56\,t\,\_Z + 49\,t\,),$$
$$x = \mathrm{RootOf}((\,79\,t^5 + 57\,t^2 - 59\,t^3\,)\,\_Z^4 + 63\,\_Z^2 + 56\,t\,\_Z + 49\,t\,)\}\}$$

Strictly speaking, that is an improvement—we could, if we so desired, write explicit formulas for the roots of that degree 4 polynomial, and thus give an explicit formulation for the curve described by that random degree 5 polynomial. However, the explicit parametric description is sufficiently complicated that we would tend to just use the original formulation.

Exercises

1. By carefully investigating the parameterization of the Folium of Descartes above, decide why the Maple plot was so rough.

2. In Chapter 2, you were asked to plot the Cissoid of Diocles, whose rectangular equation is

$$y^2 = \frac{x^3}{2a - x}$$

and whose standard parametric equations are $x = 2a\sin^2\theta$, $y = 2a\sin^3\theta/\cos\theta$. Try parsolve on this problem, and comment.

3. Try to find a transcendental equation for which this parameterization technique succeeds.

4. Write to me if you have found better or more general tricks for solving equations parametrically.

3.10.2 Reading a procedure into Maple

I switch back and forth from worksheets to ASCII use of Maple, and use Maple scripts primarily in ASCII. This means that it is sensible to put interface(echo=2) in my maple.ini file, so I get readable simulations of interactive Maple sessions when the scripts are read in. However, I occasionally write procedures, and while debugging them, I wish to read the new version in without having the input echoed. So for that read, it is best to turn echoing off. The simplest answer is to write a new procedure (called load, below) that loads in Maple procedures. This version is very simple:

```
load := proc(f);
  interface(echo=0);
  read f;
  interface(echo=2);
end:
```

Exercises

1. Modify the above procedure so it does not disturb the setting of echo.

2. Modify the above procedure so that it checks that its input is a string.

3.10.3 Indexing function for banded matrices

The following more complicated example shows how to use the new indexing functions of Maple V Release 3. For more details on this program, see (13), where this example is described together with some other programs by Khaled El-Sawy. Jérôme Lang and David Clark of the Maple group were kind enough to help make the following program work.

Before we get into the details of the program, however, it may be helpful to explain just what an 'indexing function' is. An indexing function is a procedure which accepts as input an *index* (which in the case of a matrix is a pair of integers, indicating the row and column number) and returns the corresponding matrix element location. In Maple, we need the location only to fetch the value or to store the value, and so a Maple indexing function doesn't actually return the location of the matrix entry, it just allows you to fetch or store at that location.

The following program creates an indexing function for matrices with (p, q) bandwidth (that is, p nonzero subdiagonals and q nonzero superdiagonals). It uses substitution of the global variables 'index/band/lower' and 'index/band/upper' into a *template* procedure, and assigns the result to 'index/band<p>-<q>'.

```
BandIndexFcn := proc(p:posint, q:posint);
  assign( 'index/band'.p.'-'.q,
    subs('index/band/lower'=p, 'index/band/upper'=q,
        proc(indices, tabl) local i,j;
          i := indices[1];
          j := indices[2];
          if nargs=2 then
            if type(indices, list(integer)) and
               (i-j > 'index/band/lower' or j-i > 'index/band/upper') then
              0
            else
                eval(tabl[op(indices)], 1)
            fi
          elif type(indices, list(integer)) and
               (i-j > 'index/band/lower' or j-i > 'index/band/upper') then
                 ERROR('band width', 'index/band/lower', 'index/band/upper',
                     'exceeded at', i,j)
          else
             tabl[op(indices)] := args[3..nargs]
          fi
        end));
  'indexing function band'.p.'-'.q.' created.';
  end:
```

The basic mechanism of the new indexing functions is that if they are called with three arguments, then assignment is occurring. If instead they are called with only two arguments, then the value of the array entry is wanted. Recursive use of the various possible indexing functions is automatic—the table that is passed into this indexing function has had one indexing function name (band<p>-<q>) stripped from its list, and so further references are made with respect to all remaining indexing functions, if any. If there are none, the built-in table indexing is used. This provides a convenient way to define matrices with more than one property.

Remark. A matrix A with a band indexing function defined by the procedure above will return 0 if asked to evaluate A_{ij} where i or j is outside the bounds of the matrix. We could correct this by testing every time if the indices i and j are inside the bounds. For efficiency, we prefer not to do this, since any program that refers to the matrix outside the bounds is in error anyway (though it is possible that by returning zero we will allow the program to compute correct answers).

This is in keeping with the seeming philosophy of the new indexing functions, which allow the user to get him- or herself into trouble by mixing incompatible indexing functions. For example, one can give a matrix both symmetric and antisymmetric indexing functions, and then the results of print(A) depend on the ordering of the indexing functions. We do not regard this as a bug, but rather as a useful freedom, which requires concomitant care on the part of the user.

It is useful in programs to determine from the matrix object what its bandwidth is. We use sscanf to accomplish this, by scanning the index function name.

```
bandwidth := proc(A:array) local n,idxfnc;
  idxfnc := [ linalg[indexfunc](A) ];
  idxfnc := select((p -> substring(p, 1..4) = ''.band), idxfnc);
  if idxfnc=[] then
    n := linalg[coldim](A) - 1;
    [n, n]
  else
    sscanf(idxfnc[1], 'band%d-%d')
  fi
end:
```

Finally, explicit creation of the indexing function is a chore best done automatically. If a user wants a banded matrix, she or he simply asks for one, and the following procedure will create the necessary indexing function if it has not already been done.

```
bandmatrix := proc(p:integer, q:integer) local bandidx;
  bandidx := 'band'.p.'-'.q;
  if not assigned('index/'.bandidx) then BandIndexFcn(p,q) fi;
  array(bandidx, args[3..nargs])
end:
```

Here are some examples. We first create a random 5 by 5 banded matrix (see exercises), with one subdiagonal and one superdiagonal (so the matrix is tridiagonal).

```
> A := randband(1, 1, 1..5, 1..5);
```

$$A := \begin{bmatrix} -85 & -55 & 0 & 0 & 0 \\ -37 & -35 & 97 & 0 & 0 \\ 0 & 50 & 79 & 56 & 0 \\ 0 & 0 & 49 & 63 & 57 \\ 0 & 0 & 0 & -59 & 45 \end{bmatrix}$$

We have written two simple routines for LU factorization and solution of banded linear systems, based on Algorithms 5.3.1–5.3.3 in (22). The routine bandecomp (see exercises) computes the factorization, and returns the results in a matrix of the same banded shape as A.

```
> F := bandecomp(A);
```

$$F := \begin{bmatrix} -85 & -55 & 0 & 0 & 0 \\ \dfrac{37}{85} & \dfrac{-188}{17} & 97 & 0 & 0 \\ 0 & \dfrac{-425}{94} & \dfrac{48651}{94} & 56 & 0 \\ 0 & 0 & \dfrac{4606}{48651} & \dfrac{2807077}{48651} & 57 \\ 0 & 0 & 0 & \dfrac{-2870409}{2807077} & \dfrac{289931778}{2807077} \end{bmatrix}$$

Now choose a random right-hand side.

```
> b := linalg[randvector](5);
```

$$b := [-8, -93, 92, 43, -62]$$

We can solve this system by solving $Ly = b$ and $Ux = y$ in turn. This is encoded in bandsolve.

```
> x := bandsolve(F, b);
```

$$x := \left[\frac{-97936493}{96643926}, \frac{827068483}{483219630}, \frac{-35165237}{48321963}, \frac{18382928}{16107321}, \frac{17187770}{144965889} \right]$$

We verify that the solution is correct by examining the residual.

```
> evalm(A&*x - b);
```

$$[0, 0, 0, 0, 0]$$

Maple has some advantages over purely numerical systems. For instance, one can compute the residual at a higher precision than that used for calculation, if that is necessary (here we used exact arithmetic and so the residual is exactly zero).

Another advantage is that we can experiment with the use of ∞. In the routine bandecomp, we chose to replace zero pivots with $1/\infty$. Since we are not going to exchange rows, a zero pivot normally means termination, but we thought we might see what happened if we let Maple handle this (hopefully rare) case symbolically.

```
> B := bandmatrix(1, 1, 1..2, 1..2, [ [0,1], [2,0] ]);
```

$$B := \begin{bmatrix} 0 & 1 \\ 2 & 0 \end{bmatrix}$$

This matrix obviously has a zero pivot.

```
> c := linalg[randvector](2);
```

$$c := [\, 77, \, 66 \,]$$

```
> BF := bandecomp(B);
```

$$BF := \begin{bmatrix} \dfrac{1}{\infty} & 1 \\ 2\,\infty & -2\,\infty \end{bmatrix}$$

We see that the factorization contains ∞.

```
> xb := bandsolve(BF, c);
```

$$xb := \left[\left(77 + \frac{1}{2} \frac{66 - 154\,\infty}{\infty} \right) \infty, \; -\frac{1}{2} \frac{66 - 154\,\infty}{\infty} \right]$$

```
> map(expand, ");
```

$$\left[33, \; -33\,\frac{1}{\infty} + 77 \right]$$

Now that solution can be interpreted as [33, 77], which, as it turns out, is the exact solution to the original problem. Unfortunately, the only reason that worked is because of a bug in the infinity arithmetic in Maple: it is supposed to complain if we try to cancel infinity/infinity.

```
> evalm(B &* " - c);
```

$$\left[11\frac{-3 + 7\infty}{\infty} - 77, \ 0 \right]$$

```
> map(expand, ");
```

$$\left[-33\frac{1}{\infty}, \ 0 \right]$$

This approach is equivalent to replacing a zero pivot with a small ϵ, solving the resulting system, and then taking the limit as $\epsilon \to 0$. So long as the original matrix is not singular, it should produce correct answers, provided any infinities arising on different zero pivots (supposing that there are more than one) do not cancel out. Maple's infinity arithmetic is supposed to complain if it is asked to subtract $\infty - \infty$ or divide ∞/∞, but unfortunately the implementation of these errors is not complete. Hence there is a small possibility that the results of this program will be wrong.

It may also be that processing the symbol `infinity` will add a significant overhead to the efficiency of this program. If that turns out to be the case, we should remove the statement and replace it with an error message.

We are presuming that *small* pivots are not a problem, because of the arbitrary precision floating-point arithmetic. If rounding errors are causing a difficulty, then the precision will be increased. This, too, can be very computationally expensive, but if it is too expensive, then likely this is the wrong approach anyway and some form of partial pivoting must be tried.

This indexing function approach has the advantage of being totally transparent to the user—once these indexing functions have been defined, all the utility programs such as `print` and `evalm` will do the correct things with these matrices.

Exercises

1. Count the number of rationals Maple must store in a tridiagonal n by n system. Compare it to how many it must store in the inverse. The fact that the inverse of a tridiagonal system is *full* effectively demonstrates that you should never invert a tridiagonal matrix to solve a tridiagonal system $Ax = b$.

2. Write a procedure `randband` that will generate a random banded matrix.

3. Write a procedure `bandecomp` that will factor a (p, q)-banded matrix A, without partial pivoting. Write another procedure that will use partial pivoting (note that you can bound the growth in the band width in advance).

4. Write a procedure `bandsolve` that will solve a banded system $Ax = b$ given the factored form of A and the vector b. Compare the speed of your programs with the built-in linear algebra solver. See (13) for an answer to this question.

5. Write a procedure, modeled perhaps on `'linalg/matrix'`, which will generate a banded matrix of a given dimension with entries specified by a given function $(i, j) \mapsto f(i, j)$. Test your procedure.

3.10.4 Solution of $y'(t) = ay(t-1)$

The following procedure uses a residue formula due to Wright (45) for the solution of the linear scalar delay equation $y'(t) = ay(t-1)$, with given initial function $y(t) = f(t)$ on $0 < t \le 1$, and where $y(0) = y_0$ is given. Wright's formula is more general than the one used here, which uses the Lambert W function (15). However, there are some new ideas here also, and in particular we introduce the Discrete Lambert Transform.

```
Wright := proc(t:name, a:{numeric,constant},
               f, y0:{numeric,constant}, n:posint)
  local ans,df,k,m,res,s,u,w,integralformula;

  # We use the environment variables Normalizer and Testzero
  # to tell series (which is called by residue) how to recognize
  # when w*exp(w) - a is zero.

  Normalizer := simplify;
  Testzero   := b -> evalb(Normalizer(b)=0);

  # f(t) is given.

  df := unapply(diff(f,t), t);

  w := array(-n..n):
  for k from -n to n do
    w[k] := evalf(W(k,a));
  od:

  # This is H(s) from (1.8) in Wright (1947).

  integralformula := exp(s)*y0 + exp(s) * int( df(u)*exp(-s*u), u=0..1);

  if has(integralformula, int) then
```

```
    ERROR(
    'Sorry, can't integrate to find the function to compute the residue.')
  else
    res := readlib(residue)(integralformula/(s*exp(s)-a), s=W(m,a));
    userinfo(1, Wright, cat('Residue is', res));
    ans := 0;
    for k from -n to n do
      ans := ans + subs(W(m,a)=w[k], res)*exp( w[k]*t );
    od;
    if evalc( Im( a ) ) = 0 then
      evalf( evalc( Re( ans ) ) )
    else
      evalf( ans )
    fi
  fi
end:
```

The scope of this procedure is limited to the initial functions for which the integral and residue can be calculated. The bottleneck is the integration, I suspect. As in the `Fourier_sine` procedure with which we started this book, this procedure will return an approximate answer that is a finite sum of exponential and trigonometric terms.

Of particular programming interest is the use of the environment variables `Normalizer` and `Testzero` to make sure that `residue` gets the right answer. Also, if you wish to see the residue itself, you can ask this procedure to print it by issuing the command `infolevel[Wright] := 1`.

The rest of the procedure is a straightforward implementation of the mathematics. Here is an example of its use.

```
> Wright(t, -1/10, exp(-t), 1, 5);
```

$$.02711485080\, e^{(-5.601410990\, t)} \cos(26.49519305\, t)$$
$$-.1987874468\, e^{(-5.601410990\, t)} \sin(26.49519305\, t)$$
$$+.04319168184\, e^{(-5.340259670\, t)} \cos(20.16142319\, t)$$
$$-.2591482800\, e^{(-5.340259670\, t)} \sin(20.16142319\, t)$$
$$+.08169768368\, e^{(-4.988013626\, t)} \cos(13.79009856\, t)$$
$$-.3724020294\, e^{(-4.988013626\, t)} \sin(13.79009856\, t)$$
$$+.2308744224\, e^{(-4.449098179\, t)} \cos(7.307060789\, t)$$
$$-.6629628810\, e^{(-4.449098179\, t)} \sin(7.307060789\, t)$$
$$+1.442768651\, e^{(-3.577152064\, t)} + .3797676359\, e^{(-.1118325592\, t)}$$
$$+.009380244465\, e^{(-5.808794440\, t)} \cos(32.81150306\, t)$$
$$-.08062417707\, e^{(-5.808794440\, t)} \sin(32.81150306\, t)$$

```
> plot("/exp(-t), t=0..1);
```

Figure 3.2 Relative error in the solution with five terms

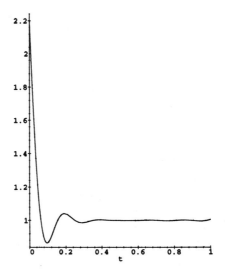

By differentiating that expression and substituting it into $y'(t) + 1/10y(t-1)$, we see that it satisfies the differential equation to within $2 \cdot 10^{-8}$. If we work to higher precision than ten digits, we can make the residual as small as we please. The plot is shown in Figure 3.2. The Gibbs phenomenon (6) is clearly visible.

The poor convergence shown can be cured by using a different approach, which I have chosen to call the Discrete Lambert Transform. Instead of trying to compute a finite number of the exact coefficients in the series

$$y(t) = \sum_{k=-\infty}^{\infty} c_k e^{W_k(a)t} ,$$

we instead compute a finite sum directly. That is, we compute

$$y(t) = \sum_{k=-N}^{N} C_k e^{W_k(a)t}$$

where the C_k are chosen not as their exact counterparts c_k in the infinite series but rather so as to minimize the L_2-norm of the difference between $y(t)$ and $f(t)$ on $0 \le t \le 1$. This means that we are solving $y'(t) = ay(t-1)$ subject to $y(t) = \hat{f}(t)$ on $0 \le t \le 1$, where $\hat{f}(t)$ is supposed to be close to $f(t)$ in the L_2-norm. This gives us the problem of minimizing

$$\int_0^1 \left(f(t) - \sum_{k=-N}^{N} C_k e^{W_k(a)t} \right)^2 dt .$$

We differentiate with respect to each C_j and set the resulting equations to zero. The solution gives a global minimum because the functional is convex. This gives us $2N + 1$ linear equations in the $2N + 1$ unknowns C_j. It is easy to see that the equations are

$$\sum_{k=-N}^{N} C_k \int_0^1 e^{(W_k(a)+W_j(a))t}\, dt = \int_0^1 f(t)e^{W_j(a)t}\, dt \qquad \text{for } j = -N, \ldots N$$

and so the matrix entries are constant on anti-diagonals $k + j$ =constant. This means the system is a Hankel system, which can be solved easily by an $O(N^2)$ algorithm, or even faster for N large. George Labahn informs me that there are $O(N \log^2(N))$ algorithms for such problems, but they begin to pay off only for N about 600 or so. The integrals in the matrix can be computed analytically—they are only exponentials after all—but the integrals on the right-hand side can only be computed once the function f is known, and perhaps they can only be done numerically.

Exercises

1. Write a procedure to find the Discrete Lambert Transform with respect to a of a given function $f(t)$. Do not worry about fast algorithms to solve the Hankel system. Redo the example above where we used residues before, and compare the accuracy of the answer with the previous one by comparing the residual on $0 \le t \le 1$. You should find that the Discrete Lambert Transform gives a more accurate answer, although the error is more evenly spread across the interval $0 \le t \le 1$.

2. For which numbers a in $y'(t) = ay(t - 1)$ does the influence of perturbations in $f(t)$ die away as $t \to \infty$?

3. For which numbers a in $y'(t) = ay(t-1)$ does the influence of persistent perturbations $y'(t) = ay(t-1) + \epsilon v(t)$ remain bounded? Take $\epsilon > 0$ and $||v|| \le 1$.

4 Keyword summary

This chapter provides a 'functionally organized' set of pointers to Maple commands. See also your help browser, if your system has one. The material presented here is expanded in the help files—the main intention of this chapter is to let you know what you need help *on*. To find further information on the topics presented here, issue the command ?<keyword> in Maple, consult the index to the Maple Library Reference Manual (10), or consult The Maple Handbook (39). You will sometimes find that there is more than one Maple routine with the name you ask for help on, and you will be asked for a more specific choice (e.g., ?basis will result in Maple asking you to try one of ?NPspinor[basis], ?linalg[basis], or simplex[basis]). Note that this keyword summary covers the Maple Share Library as well as the main Maple library.

Each section of this chapter contains the names of Maple commands relevant to that particular area: for example, the first section contains all the Maple commands relevant to linear algebra calculations. Detailed instructions for the command usage are *not* included; the intention of this chapter is that you will browse it (using the Table of Contents as an index) to find the names of the commands that do what you want, and then you will use the help facility to get more details on the usage.

In each mathematical section, you will find subsections organizing the material into logical divisions, such as 'eigenproblems' or 'univariate calculus.' In addition, you will find sections listing the relevant parts of the convert command, and the relevant Maple data types testable by the Maple structured type tests.

4.1 Linear algebra

4.1.1 Essentials

evalm . . . . Evaluation of matrix expressions (see also `&*`)

linalg . . . . The linear algebra package

matrix . . . . Define a matrix

multiply . . . . Multiply two matrices (see also `evalm` and `&*`)

transpose . . . . Take the transpose of a matrix (see also `htranspose` for Hermitian transpose)

vector . . . . Define a vector

with . . . . Loading packages

4.1.2 Solution of linear systems

addcol . . . . Form linear combinations of columns

addrow . . . . Form linear combinations of rows

adjoint . . . . Matrix of cofactors. `adj` is another name for the same routine

backsub . . . . Back substitution

basis . . . . Compute a basis for a vector space

cond . . . . Condition number (still important for exact solutions)

det . . . . Determinant

ffgausselim . . . . Fraction-free Gaussian elimination

gausselim . . . . Gaussian elimination (this is less efficient than `ffgausselim` for integers)

gaussjord . . . . Gauss-Jordan reduction (`rref` is another name for the same routine)

inverse . . . . Matrix inverse (use only when absolutely necessary)

leastsqrs . . . . Find the least squares solution

linsolve . . . . Solution of sparse systems

nullspace . . . . Null space of a matrix (see also `colspace`, `rowspace`) rank . . . . Rank of a matrix

rref . . . . Reduced Row Echelon Form (see also `RowEchelon` in the Share Library)

4.1.3 Eigenproblems

charpoly . . . . Characteristic polynomial of a matrix. See also `charmat`

definite . . . . Test if a matrix is positive, negative, semi-definite, or indefinite

eigenvals . . . . Eigenvalues of a matrix

Eigenvals . . . . Inert placeholder for invoking floating-point eigenvalue finder

eigenvects . . . . Eigenvectors of a matrix

exponential . . . . Compute the exponential of a matrix

jordan . . . . Compute the Jordan Canonical Form

minpoly . . . . Compute the minimum polynomial of a matrix

RowEchelon . . . . Row echelon decomposition or factorization of a matrix, with provisos (12, vol. 1) (see also the Share Library). Issue the commands `with(share)`; `?readshare(Echelon, linalg)`; and then `?RowEchelon`

singularvals . . . . Compute the singular values of a matrix (see also `SVD`)

Svd . . . . Inert placeholder for invoking floating-point singular-value decomposition (see also `singularvals`)

4.1.4 Orthogonalization

basis . . . . Compute a basis for a vector space

GramSchmidt . . . . Orthogonalization

orthog . . . . Test if a matrix is orthogonal

4.1.5 Special matrices

band	Create a banded matrix
bezout	Bézout matrix of two polynomials (see also `resultant` and `sylvester`)
blockmatrix	Create a block matrix
companion	Companion matrix of a polynomial p (the eigenvalues of this matrix are the roots of the given polynomial p)
diag	Generate a diagonal matrix from a vector or list
fibonacci	Fibonacci matrices
hankel	Hankel matrices (See Chapter 3, not `linalg`)
hilbert	Hilbert matrices (classical ill-conditioned matrices)
identity	`&*()` or `array(1..n,1..n,identity)` (see also `?evalm`)
randmatrix	Random matrices
sparse	Sparse matrix indexing function
sylvester	Sylvester matrix of two polynomials (see also `resultant` and `bezout`)
symmetric	Symmetric indexing function
toeplitz	Toeplitz matrices
vandermonde	Vandermonde matrices (used in interpolation theory)

4.1.6 Theory

colspace	Basis for the column space of a matrix
colspan	Same as `colspace`
frobenius	Compute the Frobenius form (rational canonical form) of a matrix (`ratform` is the same)
hadamard	Hadamard bound on the coefficients of a determinant
hermite	Hermite normal form
intbasis	Determine a basis for the intersection of spaces

minor . . . . Compute a specific minor of a matrix

norm . . . . Norm of a matrix

nullspace . . . . Basis for the null space of a matrix (`kernel` is the same thing)

permanent . . . . Compute the permanent of a matrix

rowspace . . . . Basis for the row space of a matrix

rowspan . . . . Same as `rowspace`

smith . . . . Smith normal form (see also `ismith`)

sumbasis . . . . Determine a basis for the sum of vector spaces

trace . . . . Compute the trace of a matrix

4.1.7   Utilities

augment . . . . Concatenate (join) two matrices together (same as `concat`)

col . . . . Extract a column

coldim . . . . Column dimension of a matrix

concat . . . . Concatenate (join) two matrices together (same as `augment`)

copy . . . . Copy a matrix or table

copyinto . . . . Copy into a matrix

delcols . . . . Delete columns of a matrix

delrows . . . . Delete rows of a matrix

diag . . . . Generate a diagonal matrix from a vector or list

entermatrix . . . . Prompted matrix entry

entries . . . . List the entries of a matrix

equal . . . . Test when two matrices are equal

extend . . . . Extend a matrix by rows or columns

eval . . . . Evaluate a matrix—chiefly used to look at a matrix instead of at the name of a matrix, which is the default result on entering the name of the matrix, by Maple's last-name evaluation rules

genmatrix . . . . Generate a matrix from a linear system of equations

indices . . . . List the indices of a table or array

iszero . . . . Test when a matrix is the zero matrix

mulcol . . . . Multiply a column by a scalar

mulrow . . . . Multiply a row by a scalar

op . . . . Operands of an object. Use eval for matrices instead

row . . . . Extract a row

rowdim . . . . Row dimension of matrix

stack . . . . Stack matrices to form a new matrix

submatrix . . . . Extract a submatrix from a matrix

swapcol . . . . Exchange columns

swaprow . . . . Exchange rows

vectdim . . . . Dimension of vector

4.1.8 Vector operations

angle . . . . Angle between two vectors

crossprod . . . . Cross product

dotprod . . . . Dot product

innerprod . . . . Inner product of a sequence of vectors and matrices

normalize . . . . Normalize a vector (make it a unit vector)

randvector . . . . Random vector

subvector . . . . Extract a subvector from a vector

vectdim . . . . Dimension of vector

4.1.9 Linear programming

simplex . . . . A package that contains routines for optimizing linear functionals with linear constraints by the simplex method. The algorithm may be used as a 'black box' or step by step

4.1.10   Appropriate conversions

Use the `convert` command to convert to the following data types. For example, the entry `list` in the table below indicates that the command `convert(<blah>, list)` will do something to `<blah>`, namely, convert it to a list. For more details on what `<blah>` is allowed to be, consult `?convert[list]`.

array . . . . Convert to an array. There is no help file for this, but the function exists. If you have a list of lists, `convert(",array)` will do the conversion

list . . . . Convert to a list

matrix . . . . Convert to a matrix

std . . . . From the simplex package—conversion to standard form

stdle . . . . From the simplex package—conversion to nonstrict inequalities

table . . . . Convert to a table

vector . . . . Convert to a vector

4.1.11   Relevant structured data types

This section discusses the data types useful in linear algebra. One can test if a given object is of type `<blah>`, where `<blah>` is one of the entries below, by issuing the command `type(<object>,<blah>)`. This type-test will return `true`, `false`, or `FAIL`. The forward quotes in the entries are to prevent possible values of the keywords from invalidating the type test. For example, if one executes the commands

```
matrix := array(1..2,1..2, [[1,2],[3,1]]):
<some other commands>
type(A, matrix);
```

then you will be asking if A is of type `array(1..2,1..2, [[1,2], [3,1]])`, which will yield an error message. If, however, you issue the command

```
type(A, 'matrix');
```

then Maple will test if A is a matrix. Thus, if you *know* you have not assigned a value to `matrix` (since assigning a value to a Maple keyword is a bad idea anyway), then you may omit the quotes in what follows. For help on a given data type, issue the command ?type[<blah>]. For help on defining your own data types, issue the command ?type.

'array'.................. Equivalent to type(<object>,'array'('anything'))

'array'(K) .... Array of objects of type K

'matrix' .... Equivalent to type(<object>,'matrix'('anything'))

'matrix'(K) .... Matrix of objects of type K

'matrix'(K, square) ...... Square matrix of objects of type K

'vector'................. Equivalent to 'vector'('anything')

'vector'(K).............. Vector of objects of type K

4.2   Polynomials and rational functions

There is a good description of polynomials available through ?polynomial and a similar description for rational functions through ?ratpoly.

4.2.1   Polynomial manipulations

asubs................... Substitute expressions into a sum using semantic substitution (see also subs)

coeff................... Pick off coefficients of an *expanded* or *collected* polynomial (see also coeffs and coeftayl)

collect................. Collect like terms, and optionally apply a function (such as factor) to each coefficient

compoly................. Write a polynomial as a composition of two polynomials, if possible

eliminate .... Eliminate some variables from a set of equations

expand.................. Expand a product of polynomials

factor .... Factor a (multivariate) polynomial (*very useful*) (see also ifactor to factorize an integer, and factors)

factors .... Compute a list of the factors of a polynomial

horner .... Convert a polynomial to Horner form

irreduc................. Decide if a polynomial is irreducible

sort	Sort a polynomial into a sensible order
split	Factor a polynomial completely, into linear factors
sqrfree	Factor a polynomial into square-free factors
subs	Substitute into an expression
unapply	Convert an expression to an operator (function)

4.2.2   Mathematical operations

+, −, *, /, ∧	Arithmetic operators for addition, subtraction, multiplication, division, and raising to a power
content	Determine the content of a polynomial
diff	Differentiate a function
discrim	Polynomial discriminant (see (12, vol. 1))
divide	Exact polynomial division
gcd	Polynomial GCD (greatest common divisor)
gcdex	Find the GCD d of a and b by the Extended Euclidean Algorithm, so the polynomials s and t such that $d = s \cdot a + t \cdot b$ can be returned
int	Integrate a function
lcm	Polynomial LCM (least common multiple)
prem	Pseudo-remainder
proot	n^{th}-root of a polynomial
psqrt	Square root of a polynomial, if it exists
rem	Remainder
resultant	Polynomial resultant (see (12, vol. 1), and also bezout and sylvester)
sprem	Sparse pseudo-remainder
sum	Find an anti-difference of an expression
quo	Quotient

4.2.3 Rootfinding

companion	Create the companion matrix of a polynomial. Its eigenvalues are the roots
fsolve	Numerically find roots of polynomials or systems of polynomials
galois	Compute the Galois group of an irreducible polynomial of degree up to 7 (see also GF, the Galois Field package)
grobner	Package for computing Gröbner bases
gsolve	Solve a system of polynomial equations via a Gröbner basis
isolate	Isolate a term in an algebraic equation
realroot	Find dyadic rational guaranteed bounding intervals for the real roots of a polynomial
RootOf	Symbolic representation of a generic root of a polynomial
roots	Exact roots of a univariate polynomial over an algebraic number field
Roots	Roots of a polynomial mod n
solve	Solve a polynomial system (see also grobner)
sturm	Find the number of real roots in a given interval by use of Stürm sequences

4.2.4 Symbolic manipulation of roots

alias	Use short names (e.g., for bulky RootOfs)
allvalues	Compute radical values of a RootOf expression. The d option is often what is wanted (see also convert,RootOf and convert,radical)
evala	Evaluate over an algebraic field (manipulate RootOfs)
Factor	Factor a polynomial with algebraic coefficients
Gcd	Gcd of polynomials with algebraic coefficients
RootOf	Symbolic representation of a generic root of a polynomial

roots . . . . Exact roots of a univariate polynomial over an algebraic field

4.2.5 Orthogonal polynomials

convert . . . . To convert a Chebyshev series to Padé form, use the same command to convert Taylor series to Padé form, i.e., `convert(p, ratpoly);`

orthopoly . . . . The orthogonal polynomial package: definitions of Gegenbauer, Hermite, Laguerre, generalized Laguerre, Legendre, Jacobi, and Chebyshev polynomials of the first and second kind

4.2.6 Rational functions

convert . . . . Convert to continued fraction form, partial fraction form, or Padé form (see `?convert` for the precise syntaxes of each of these conversions)

FPS . . . . (Share Library) Formal power series. Do `with(share); readshare(FPS, calculus);` and then `?FPS`

fullparfrac . . . . (Share Library) Full partial fraction decomposition. Do `with(share); readshare (fparfrac, calculus);` and then `?fullparfrac`

normal . . . . Cancel common factors

residue . . . . Find the residue at a pole

series . . . . Compute a Laurent (or more general) series

4.2.7 Interpolation and approximation

bernstein . . . . Bernstein approximating polynomial

chebyshev . . . . Chebyshev series expansion of a function

dinterp . . . . Determine probabilistically the degree of an expression in a given variable

FFT . . . . Compute the pure-power-of-two Fast Fourier Transform of a set of data

iFFT	Inverse Fast Fourier Transform
interp	Polynomial interpolation
leastsqrs	(`linalg` package) Least squares approximation (see also `vandermonde`)
piecewise	Piecewise function definition
ratrecon	Reconstruction of a rational polynomial from its images modulo a polynomial
series	Compute series approximations
sinterp	Sparse probabilistic interpolation
spline	Spline approximation (see also `bspline`)
taylor	Compute Taylor polynomial approximations
thiele	Thiele rational interpolation

4.2.8 Utilities

degree	Compute the degree of a polynomial
lcoeff	Compute the leading coefficient (coefficient of the highest-degree term)
ldegree	Compute the degree of the lowest-degree term of the polynomial
maxnorm	Compute the maximum of the absolute values of the coefficients of the polynomial
norm	Compute various algebraic norms of a polynomial
op	Examine the operands of a polynomial (or indeed of any Maple object)
recipoly	Test if a polynomial is self-reciprocal, i.e., if the roots occur in reciprocal pairs
sign	Compute the signum of the leading coefficient of a polynomial. Note that a common Maple error is to use `sign` when you mean `signum`
tcoeff	Trailing coefficient of a polynomial
translate	Change x to $x + a$—that is, shift the origin of a polynomial

4.2.9 Appropriate conversions

For help on each of these items, issue the command `?convert[<item>]`.

'+' .... Convert a list of terms to a sum—that is, add up the terms in a list. There is no help file for this. See section 2.4 instead

''*'' .... Convert a list of terms to a product—that is, multiply out the terms in a list. See section 2.4 since there is no help file for this

confrac .... Convert a rational function to continued-fraction form

horner .... Convert to Horner form for rapid evaluation

mathorner .... Convert to matrix Horner form for polynomials of matrices

parfrac .... Convert to partial fraction form—see also `fullparfrac` in the Share Library

polynom .... Convert a series object to a sum of terms

radical .... Convert `RootOf`'s to radicals, if possible

ratpoly .... Convert a series object to Padé form

RootOf .... Convert radicals to `RootOf`'s

sqrfree .... Convert to square-free factored form

4.2.10 Relevant structured data types

For help on a given data type, issue the command `?type[<blah>]`. For help on defining your own data types, issue the command `?type`. Note that the linear, quadratic, cubic, and quartic types do not recognize degenerate cases. For example, numbers are not of type `linear`.

'+' .... A sum of terms of any type (similarly for '*' and the other operators)

'algebraic' .... Any single mathematical expression, including transcendental functions

'anything' .... Every Maple object is of type `anything`, except expression sequences

'cubic' .... A cubic polynomial

'expanded' .... An expanded expression

'laurent' .... Laurent series

'linear'	A linear polynomial
'monomial'	A single term in one or more variables
'polynom'	Equivalent to `type(<object>,'polynom'('anything'))`
'polynom'(K, variable) . . .	True if the object is a polynomial in the variable 'variable' with all its coefficients of type K
'quadratic'	Quadratic polynomial (parabola)
'quartic'	Quartic (biquadratic) or fourth-degree polynomial
'radical'	An expression containing a root extraction
'ratpoly'	A ratio of two polynomials, or rational function
'RootOf'	A `RootOf` expression
'series'	A series object
'sqrt'	A square root of something
'taylor'	A pure Taylor series

4.3 Calculus

4.3.1 Univariate calculus

asympt	Compute an asymptotic series
D	Differentiate an operator (see also `Diff`, `diff`)
diff	Differentiate an expression (see also `Diff`, `D`)
DESol	Symbolic representation of the solution of a differential equation
DEtools	Package of tools for manipulating differential equations, including change of variables, phase plots, and so on
dsolve	Solve differential equations
gfun	(In the Share Library) Generating function manipulations. Do `with(share);` `readshare (gfun, calculus);` and then `?gfun`
int	Integrate an expression. See also `Int` (the inert form) and `?evalf[int]`

laplace . . . . Compute Laplace transforms (see also `fourier` and `mellin`)

limit . . . . Compute the limit of an expression

series . . . . Compute a finite segment of a (generalized) series (see also `powseries` and the Share Library power series package FPS, which contains a facility for computing infinite series)

student . . . . The `student` package contains many useful utilities for calculus, including change of variables (`changevar`) and integration by parts (`intparts`)

taylor . . . . Compute a Taylor series (restricted version of `series`)

Wronskian . . . . (In the `linalg` package) Compute the Wronskian determinant

4.3.2   Multivariate calculus

curl . . . . (In the `linalg` package) Curl of a function

D . . . . Compute partial derivatives of operators

diff . . . . Compute partial derivatives of expressions

diverge . . . . (In the `linalg` package) Compute the divergence

DoubleInt . . . . (In the `student` package) Easy interface for multiple integration

grad . . . . (In the `linalg` package) Gradient of a vector function in Cartesian, spherical, cylindrical, or orthogonally curvilinear coordinates

hessian . . . . (In the `linalg` package) Compute the Hessian matrix of second derivatives of a multivariate function

jacobian . . . . (In the `linalg` package) Compute the matrix of derivatives of a vector function

laplacian . . . . (In the `linalg` package) Compute the Laplacian operator in Cartesian, spherical, cylindrical, or orthogonally curvilinear coordinates liesymm . . . . The Lie Symmetries package for exact solution of partial differential equations

mixpar . . . . Make sure mixed partial derivatives commute

mtaylor . . . . Compute the multivariate Taylor series

PDEplot . . . . In the DEtools package. Plot the solution of a certain class of PDEs

potential . . . . (In the linalg package) Compute the potential of a vector field

vecpotent . . . . (In the linalg package) Compute the vector potential

4.3.3 Complex variables

abs . . . . Absolute value or modulus

argument . . . . Argument or phase angle

assume . . . . Assume an object has properties—e.g., that it is complex

conjugate . . . . Complex conjugate of a complex number

csgn . . . . Complex signum function, which makes it easy to encode information about closure on branch cuts. A very useful function, unique to Maple (so far)

evalc . . . . Evaluate expressions over the complex number field

evalf . . . . Evaluate an expression to (possibly complex) floating point

GaussInt . . . . The Gaussian Integers package

I . . . . The square root of minus one

Im . . . . Imaginary part of an expression—see evalc

polar . . . . Polar form of a complex number

Re . . . . Real part of an expression—see evalc

residue . . . . Compute the residue of an expression

series . . . . Compute a Laurent series of an expression signum.................... +1 if argument is positive, −1 if argument is negative, and *user-settable* by the environment variable _Envsignum0 at 0—that is, you can decide what the signum of zero is. By default it is undefined, in a convenient way—signum(x) will return 1 if x has been assumed nonnegative, and −1 if x has been assumed nonpositive. signum(0) will return 0. For complex arguments, it returns $\exp(i\theta)$ where θ is the phase

4.3.4 Elementary and special functions and constants

See ?inifcns for a list of initially known functions in Maple including a description of the trigonometric and hyperbolic functions. Additional functions are defined in the orthogonal polynomial package orthopoly.

abs..................... Absolute value of real or complex argument

Ai...................... Airy function (see also Bi and Bessel)

arctan.................. Arctangent function. With two arguments, arctan(y,x) gives the angle subtended by the point (x, y) from the x-axis. It returns an angle in the range $-\pi < \theta \le \pi$. It is similar to FORTRAN's arctan2 function

argument............... Argument (phase) of a complex number or expression

bernoulli............... Bernoulli numbers and polynomials

BesselI................. Modified Bessel function of the first kind $I_\nu(x)$

BesselJ................. Bessel function of the first kind $J_\nu(x)$

BesselK................. Modified Bessel function of the second kind $K_\nu(x)$

BesselY................. Bessel function of the second kind $Y_\nu(x)$

Beta.................... Beta function: $\beta(x, y) = \Gamma(x)\Gamma(y)/\Gamma(x + y)$

Bi...................... Airy function (see also Ai and Bessel)

binomial................ Binomial coefficients: $\text{binomial}(n, r) = C_r^n = n!/(r!(n - r)!)$

Catalan................. Catalan's constant

ceil....................	Compute the nearest integer not less than the input
Chi	Hyperbolic cosine integral
Ci.....................	Cosine integral
constants...............	List of globally defined constants
csgn...................	Complex signum function, which makes it easy to encode information about closure on branch cuts. A very useful function, unique to Maple (so far)
dilog..................	Dilogarithm integral $= \int_1^x \ln(t)/(1-t)\,dt$
E.....................	The base of the natural logarithms. E is Maple's notation for $e = 2.7182818284\ldots$, and is implemented via an alias to exp(1)
Ei....................	Exponential integral $= \int_{-\infty}^x \exp(t)/t\,dt$
erf	Error function $= 2/\sqrt{\pi} \int_0^x \exp(-t^2)\,dt$
erfc	Complementary error function $= 1 - \mathrm{erf}(x)$
euler	Euler numbers and polynomials
exp	Exponential function: $\exp(x) = \sum_{i=0}^{\infty} x^i/i!$
factorial...............	Factorial function $\mathrm{factorial}(n) = n!$
FAIL	The boolean value FAIL, which is not true or false—though 'if' statements treat it as false
false...................	The boolean value 'false'
floor	Compute the nearest integer not greater than the input
frac	Fractional part of a real number
FresnelC	Fresnel cosine integral $= \int_0^x \cos(\pi t^2/2)\,dt$ (see also inifcns for auxiliary Fresnel integrals)
FresnelS...............	Fresnel sine integral $= \int_0^x \sin(\pi t^2/2)\,dt$ (see also inifcns for auxiliary Fresnel integrals)
gamma	The Euler-Mascheroni constant $\gamma = 0.577\ldots$
GAMMA	The Γ function $\Gamma(x) = \int_0^{\infty} t^{x-1} \exp(-t)\,dt$, and the incomplete Gamma function $\Gamma(x,a) = \int_a^{\infty} \exp(-t)t^{x-1}\,dt$
harmonic...............	The function $\mathrm{harmonic}(n) = \sum_{i=1}^n 1/i = \psi(n+1) + \gamma$

hypergeom	The generalized hypergeometric function
I	The square root of minus one
ilog10	Integer logarithm to the base 10—e.g., one less than the number of decimal digits of a natural number (see also `ilog`)
infinity	Positive real infinity, except when it means complex infinity
inifcns	List of initially known functions
Legendre	LegendreF, —E, —Pi, —Kc, —Ec, —Pic, —Kc1, —Ec1, —Pic1: Legendre's elliptic integrals of the first, second, and third kinds, in various flavours. For Legendre *polynomials* see the `orthopoly` package
Li	Logarithmic integral
ln	Natural logarithm (logarithm with base $e = 2.71828...$)
lnGAMMA	Natural logarithm of the Γ function
log	Logarithm to arbitrary base (with index—without index, means natural log)
log10	Logarithm to the base 10
max	Maximum of a list of real values
MeijerG	Special case of the Meijer G function
min	Minimum of a list of real values
O	Used to specify the order term for series (see also `series`, `asympt`, and `taylor`)
Pi	The ratio of the circumference of a circle to its diameter
Psi	Polygamma function $\psi(x) = \Gamma'(x)/\Gamma(x)$, its derivatives $\mathrm{Psi}(n,x) = \psi^{(n)}(x) = d^n\psi(x)/dx^n$
RootOf	Function for expressing roots of algebraic expressions
round	Round a float or rational to an integer
Shi	Hyperbolic sine integral
Si	Sine integral
sign	*Not* the signum of a real number. This function is for polynomials. Returns the sign of the leading coefficient

signum .... +1 if argument is positive, -1 if argument is negative, and *user-settable* by the environment variable `_Envsignum0` at 0—that is, you can decide what the signum of zero is. By default it is undefined, in a convenient way—`signum(x)` will return 1 if x has been assumed nonnegative, and -1 if x has been assumed nonpositive. `signum(0)` will return 0. For complex arguments, it returns $\exp(i\theta)$ where θ is the phase

sqrt .... Square root (often better as `x^(1/2)`)

true .... The boolean value 'true'

trunc .... Truncate a float or rational to an integer

W .... The Lambert W function, which satisfies $W(x)\exp(W(x)) = x$ (complex branches also implemented)

Zeta .... Riemann zeta function $\zeta(s) = \sum_{k=1}^{\infty} 1/k^s$ and its derivatives Zeta(n,s)= $\zeta^{(n)}(s)$. With three arguments Zeta(n,s,q)= $\zeta_q^{(n)}(s)$ where $\zeta_q(s) = \sum_{k=1}^{\infty} 1/(k+q)^s$

4.3.5 Functions with jump discontinuities

ceil .... Compute the nearest integer not less than the input

csgn .... Complex signum function, which makes it easy to encode information about closure on branch cuts. A very useful function, unique to Maple (so far)

Dirac .... Dirac delta function—see also `laplace` and `fourier`

discont .... Test if a function is discontinuous

floor .... Compute the nearest integer not greater than the input

frac .... Fractional part of a real number

Heaviside .... Heaviside unit step function

round .... Round a float or rational to an integer signum .... +1 if argument is positive, -1 if argument is negative, and *user-settable* by the environment variable `_Envsignum0` at 0—that is, you can decide what the signum of zero is. By default it is undefined, in a convenient way—`signum(x)` will return 1 if x has been assumed nonnegative, and -1 if x has been assumed nonpositive. `signum(0)` will return 0. For complex arguments, it returns $\exp(i\theta)$ where θ is the phase

trunc .... Truncate a float or rational to an integer

4.3.6 Appropriate conversions

For help on each of these items, ask for help on `convert,<item>`.

Ei .... Convert an expression containing trig, hyperbolic, or logarithmic integrals to one containing exponential integrals

erf .... Convert the complementary error function `erfc` to equivalent `erf` expression, using $\mathrm{erf}(x) = 1 - \mathrm{erfc}(x)$

erfc .... Opposite of `convert(<blah>,erf)`

exp .... Trig functions and hyperbolic functions to exponential form

expln .... Trig functions to exponential and inverse trig to logarithmic form

expsincos .... Trig functions to sin and cos form, hyperbolic functions to exponential form

GAMMA .... Factorials to GAMMA functions

ln .... Inverse trigs to logarithmic form

sincos .... Trig functions to sin and cos form, hyperbolic functions to sinh and cosh form

tan .... Convert trig functions to half-angle tangent form

trig .... Exponential functions to equivalent trig functions

4.3.7 Appropriate simplifications

One can use commands like `simplify(<object>,<restriction>)`, where `<restriction>` is one of the following entries; then only the indicated type of simplification will occur.

Ei . . . .   Simplify only with respect to `Ei`

GAMMA . . . .   Simplify only with respect to Γ

hypergeom . . . .   Simplify only with respect to `hypergeom`

ln . . . .   Simplify only with respect to `ln`

sqrt . . . .   Simplify only with respect to `sqrt`

trig . . . .   Simplify only with respect to `trig`

Similarly, one can issue the command `combine(<object>,<selection>)`, where `<selection>` is one of

exp . . . .   Make `exp(x)^2` go to `exp(2*x)`

ln . . . .   Make `ln(a)+ln(b)` go to `ln(a*b)`

power . . . .   Make $x^a y^a$ go to $(xy)^a$

trig . . . .   This one is the most useful—it replaces products and powers of sines and cosines with simple cosines and sines: for example, `cos(x)^3` is replaced by `3/4*cos(x) + 1/4*cos(3*x)`

Psi . . . .   Similar simplifications with $\psi(x) = \Gamma'(x)/\Gamma(x)$

Interestingly, if we issue commands of the form `expand(<object>, <subexpression>)` this means that the subexpressions are *not* expanded; in some sense `expand` behaves in the opposite way from `combine`.

4.3.8 Relevant structured data types

For help on a given data type, issue the command `?type[<blah>]`. For help on defining your own data types, issue the command `?type`.

'algebraic' . . . .   An expression

'algfun' . . . .   An algebraic function: a root of a polynomial

'arctrig' . . . .   Inverse trig function

'evenfunc' . . . .   Even function

'function' . . . .   Unevaluated function call

'mathfunc' . . . .   Initially known math function 'oddfunc' . . . . Odd function

'polynom' . . . . Polynomial

'radfun' . . . . Radical function

'radfunext' . . . . Radical function extension

'ratpoly' . . . . Ratio of polynomials

'sqrt' . . . . Square root

'trig' . . . . Trig function

4.4 Abstract algebra

The facilities in Maple for abstract algebra are many and varied. I rarely use them directly, hence there are few examples in this book. However, the following lists may serve as a starting point for your investigations.

4.4.1 Groups

define . . . . Define a group, ring, or other algebraic object

galois . . . . Galois group of a polynomial of degree less than or equal to 7

group . . . . The group theory package, which includes tools for computation of permutation groups given their generators, p-Sylow subgroups, orbits, centers, and others

4.4.2 Rings

GaussInt . . . . Package for performing arithmetic with the Gaussian integers

mod . . . . Arithmetic modulo n

modp1 . . . . Arithmetic modulo n, done efficiently. Calculations that can be done with this routine include `Degree`, `Det`, `Gausselim`, `Gaussjord`, `Multiply`, `Power`, and others

msolve . . . . Solve equations modulo m

4.4.3 Fields

evalgf .... Compute in an algebraic extension of a finite field

GF .... Perform arithmetic in a finite Galois Field

mod.................... Compute in $\mathbf{Z}_p$

modp1 .... Arithmetic modulo p, done efficiently. Calculations that can be done with this routine include Degree, Det, Gausselim, Gaussjord, Multiply, Power, and others

4.4.4 General domains

Gauss .... The Gauss package for parameterized domain computation

4.5 Combinatorics, number theory, and graph theory

binomial .... binomial coefficients: $\mathrm{binomial}(n, r) = C_r^n = n!/(r!(n - r)!)$

combinat............... The combinatorics package. This package includes routines for counting permutations and combinations, the number of distinct partitions of an integer (and the partitions explicitly), Bell numbers, Stirling numbers of the first and second kinds, codes, and more

ifactor.................. Factor an integer

igcd .... Compute the greatest common divisor of two or more integers

igcdex.................. Compute a gcd d of integers a and b using the Extended Euclidean algorithm, so the integers s and t such that $d = s \cdot a + t \cdot b$ can be returned

isolve .... Solve Diophantine equations

isprime................. Return true if a number is prime

networks .... The networks package. This package includes routines for creating labeled, directed, or undirected graphs, chromatic polynomials, components of the graph, trees, flows, cycles, and more

numtheory .... The number theory package, which includes Euler's totient function ('phi' function phi), the Jacobi symbol, ithprime for computing the ith prime, and others

padic .... Package for working with p-adic numbers

4.5.1 Appropriate conversions

For help on each of these items, issue the command ?convert[<item>].

base .... Convert to different bases

binary .... Convert to binary (integers only)

confrac .... Convert a rational or quadratic irrational to continued fraction form

decimal .... Convert to decimal

float .... Convert to float—equivalent to evalf

hex .... Convert to hexadecimal

octal .... Convert to octal

radical .... Convert an algebraic number to radical form, if possible

rational .... Convert a floating-point number to an equivalent rational number, using continued fractions to find the 'nicest' rational number nearby. If exact conversion is required, use convert(<num>, rational, exact)

RootOf .... Convert a radical expression to RootOf form

4.5.2 Relevant structured data types

For help on a given data type, issue the command ?type[<blah>]. For help on defining your own data types, issue the command ?type.

'algnum' .... Algebraic number

'algnumext' . . . . Algebraic extension—e.g., a non-rational algebraic number

'complex' . . . . Complex number

'complex'('integer') . . . . . . Gaussian integer

'const' . . . . Included here to avoid any confusion with constant, below. An object is of type const if it is a *differential form* of constant type. You probably want 'constant' or 'realcons'

'constant' . . . . An expression is a constant (integer, fraction, or float) (see also type,realcons and type,numeric)

'facint' . . . . Factored integer

'float' . . . . Floating-point number. Note: zero is not a floating-point number

'fraction' . . . . Rational number that is not an integer

'integer' . . . . Integer

'negint' . . . . Negative integer

'nonneg' . . . . Non-negative numeric

'nonnegint' . . . . Non-negative integer

'numeric' . . . . Integer, fraction, or float

'posint' . . . . Positive integer

'positive' . . . . Positive numeric

'primeint' . . . . Prime integer

'radext' . . . . Radical extension

'radfun' . . . . Radical function

'radfunext' . . . . Radical function extension

'radical' . . . . Radical expression

'radnum' . . . . Radical number

'radnumext' . . . . Radical number extension

'rational' . . . . Integer or fraction

'realcons' . . . . An object is of type 'realcons' if it is of type constant and the result of applying evalf@evalc to the object is of type numeric. ±infinity are also of type 'realcons'

'RootOf' . . . . RootOf expression

'sqrt' . . . . Square root (often better expressed as x^(1/2))

4.6 Geometry

There are three independent geometry packages in Maple. I have never used them; I merely report their existence.

geom3d .... Three-dimensional geometry package. Manipulations include creation of points, lines, planes, spheres, and transformations

geometry .... Two-dimensional geometry package. Manipulations apparently include creation of points, triangles, centroids, and transformations

projgeom .... The projective geometry package. Manipulations apparently include creation of points, lines, conics, and transformations

4.7 Plotting

DEtools .... DEtools package for manipulating differential equations. The plotting utilities from here include `DEplot`, `DEplot1`, `DEplot2`, `dfieldplot`, `PDEplot`, and `phaseportrait`

plot .... Two-dimensional function, data, parametric, and polar plots

plots .... The `plots` package, for more sophisticated plots, including contour and implicit plots, odeplots, display of multiple plots already created, animation, and more

plotsetup .... Initialization for plotting

plot3d .... Three-dimensional surface plots

statplot .... From the statistics package. It plots data and a linear regression fit on the same graph

4.7.1 Relevant structured data types

For help on a given data type, issue the command `?type[<blah>]`. For help on defining your own data types, issue the command `?type`.

'PLOT' .... Two-dimensional plot data structure

'PLOT3D' .... Three-dimensional plot data structure

4.8 Statistics

binomial .... binomial coefficients: binomial$(n, r) = C_r^n = n!/(r!(n-r)!)$

stats .... The statistics package. This includes routines for data entry, low-order statistics such as mean, median, mode and variance, correlations, covariance, various distributions, linear regression, and various statistical tests. The package has been completely rewritten for Maple V Release 3

4.9 Share Library

index .... index of all routines in the share library. Do with(share);, ?share,index

location .... You may already have the entire Share Library—type with(share); to see. If not, the Share Library is available by anonymous ftp to daisy.uwaterloo.ca in the directory maple, or by anonymous ftp to neptune.inf.ethz.ch. Alternatively, it is available by mail server. To access this server, send a mail message containing the text send info to maple-netlib@can.nl. Instructions on how to obtain code and a list of the contents will be mailed back to you

readshare .... A utility routine for reading individual packages from the share library, accessible once with(share); has been executed

with .... Issuing the command with(share) may load the complete Share Library (the version as of when your version of Maple was obtained) or it may not, if the system managers did not choose to include it in the installation

4.10 Programming

4.10.1 Control structures

break	Exit a loop *now*
ERROR	Return from a procedure *now*, with an error message
for	Repeat a sequence of statements a specified number of times
if	Conditionally execute a statement
goto	There is no 'goto' statement in Maple V Release 2 or earlier. There *may* be one in future releases, to allow simpler conversion of old-style FORTRAN programs into Maple
map	Map a function onto a complex object, such as a list or set
next	Terminate the current iteration of a loop; retest to see if the program should proceed to the next iteration
parse	Parse a string, converting it to Maple code, prior to executing it
while	Repeat a sequence of statements while a boolean expression is true

4.10.2 Object manipulation

evalapply	User-defined control over function application
has	Test if an object has a particular subobject in it
map	Map a function onto subparts of an object
match	A pattern matching utility. If you can, use the four-argument form of `collect`, or `select`, or anything else, rather than use `match`, as it uses `solve` and can be inefficient
op	Pick off selected operands
searchtext	Search a string for occurrence of text. See also the case-sensitive version, `SearchText`. These replace the older `search`

select	Select desired objects from a list or set
seq	Create a sequence of objects
zip	Mash two lists together according to some specified rule

4.10.3 Procedures

args	The actual arguments to a procedure (args is an expression sequence) (see also nargs)
Environment variables	Global variables with special properties
global	New to Maple V Release 3—declare a variable global
local	Declare a variable local
operators	Operators are simple procedures, that can be entered with 'angle bracket' syntax, 'arrow' syntax, or 'option operator' syntax. They correspond most closely with ordinary mathematical functions
options	Maple procedures may have options such as copyright, remember, system, operator, builtin, or others, or several combinations of these. Alternatively, they may have no options at all
proc	Maple procedure bodies begin with proc and end with end. They may recursively call themselves, have a variable number of arguments, and always return a single value (which may be NULL or a list, array, or expression sequence). Creation of a Maple procedure usually involves assigning the constructed procedure body to a name, but is not always necessary
RETURN	Return from a procedure *now*. Note the usual method for returning from a procedure is to 'fall off the bottom,' in which case the value returned is the value of the last statement executed
TEXT	The type of object Maple help entries are contained in

userinfo . . . . A mechanism for letting the user know what is going on inside the `proc`, if he or she wants to know (see also `infolevel`)

4.10.4   Debugging tools

lprint. . . . . The humble line print statement. Put some in for debugging purposes, and comment them out when you are finished

mint . . . . Syntax-checking utility (external to Maple). Checks for common mistakes, such as non-declaration of local variables, conflicts with Maple names, etc.

print . . . . Like `lprint` but prettier

printlevel . . . . Automatically print various amounts of in-formation while a procedure is executing

trace . . . . Trace the execution of a `proc` (not to be confused with `linalg[trace]`)

4.10.5   Program generation tools

freeze . . . . 'Freeze' a term, making it invulnerable to transformations

frontend. . . . . 'Freeze' all the subexpressions in an expres-sion

optimize/makeproc . . . . . . Part of `optimize`, which attempts to pro-duce an optimal straight-line program for the evaluation of an expression. `'optimize/makeproc'` will make a Maple procedure out of the straight-line program or compu-tation sequence

parse . . . . Parse a string, converting it to Maple code, prior to executing it

procbody. . . . . Create a neutralized form of the `proc` it is given, so transformations can be made on it

procmake . . . . Take a neutralized procedure and make a real `proc` out of it subs.................... Substitute expressions into other expressions or procedures. This can be used to create new procedures from preexisting templates

thaw .... Undo the effects of 'freeze'

4.10.6 Hacker's corner

For an interesting article describing a nontrivial example of these, see (25).

addressof............... Return the internal address of a Maple object

assemble .... Assemble a Maple object from component parts

disassemble.............. Disassemble a Maple object into component parts

pointto.................. Return the Maple object pointed to by an address

4.10.7 Appropriate conversions

For help on each of these items, issue the command `?convert[<item>]`.

D .... Convert occurrences of `diff` or `Diff` to `D` (operator) format

diff..................... Convert occurrences of `D` or `Diff` to `diff` (expression) format

double .... Convert to C doubles

equality .... Convert inequalities to equalities (see also `convert[lessthan]`, `convert[lessequal]`)

name................... Convert to a string—see also `convert[string]` and `parse`, which converts a Maple string to something Maple can execute

set .... Convert a list to a set—this eliminates duplicate entries

4.10.8 Relevant structured data types

It is in programming that the structured data types really shine, particularly for error-checking of arguments to procedures. One includes the type of the argument to a procedure in the procedure body, as in `proc(x:matrix(integer,square))` which will accept only a square matrix with integer entries as actual argument to the procedure. So, in some sense, *all* the structured data types are relevant here. There are nearly 100 *simple* data types listed in the Maple Handbook (39), and they can each be used in combination with others to define arbitrarily complex data types. A complete list of simple types is available through `?type`. You may also define your own data types. See also the Maple Language Reference Manual (9).

4.11 File I/O

The following commands are useful for reading and writing data to and from files.

appendto	Add Maple output to an existing file
C	Write C code: `C(blah, filename='f.out')` appends the C code to the file `f.out`
echo	In your `.mapleinit` file, I recommend that you put the line `interface(echo=2)`. This ensures that when files containing Maple commands are read in, the command is echoed together with its output. See the discussion in section 1.3
fortran	Write Fortran code: `fortran(blah, filename='f.out')` appends the FORTRAN code fragment to the file `f.out`
hostfile	The command `convert(<unixname>, hostfile)` will convert a Unix-style filename to the host computer file format
interface	Many interface variables control width of the screen, terminal type for graphics output, plot devices, and the like
libname	Set or interrogate the location of the Maple library. Use of multiple libraries can be handy

plotoutput.............. Command `interface(plotoutput=<file>)` causes output of the subsequent graph to the specified file

printf .... Formatted print statement

read.................... Read a file of Maple commands. This command is very, very useful, but make sure `echo` is set according to your preference

readdata .... An attempt at automatic detection of the format for data in a file; Maple will try to read all the data from the specified file, interpreting columns of data appropriately

readline .... Read a line from the interactive terminal (see also `readstat`)

readstat .... Read a statement from the interactive terminal. Use of this statement is discouraged by the Maple group, not because the command doesn't work, but rather because it is felt that asking the user questions and expecting answers is bad programming style

save.................... Save Maple results either in human-and-machine readable form, or in machine-readable form

sharename.............. The location of the Maple Share Library

sscanf .... Formatted read statement

writeto .... Send Maple output to a file, or back to the terminal

4.12 Connections to other programs and languages

4.12.1 Conversion to FORTRAN and C

This is done with the `fortran` and `C` commands. Each of these must be loaded in via `readlib`. Conversions to other languages, such as MATLAB, are planned.

4.12.2 Typesetting language output

Maple can convert its expressions to LaTeX, TeX, or $\mathcal{AMS}$-TeX; (for the latter two, you can use the `tex` package from the Share Library, while latex conversion is built in to Maple). In addition, conversion to *troff/eqn* is available via the eqn command.

4.12.3 Calling other programs from Maple

Currently the only way to do this is via the system command. You must write out the input to this program to a file, call the program, and then read the results back in from a file for further processing.

4.12.4 Calling Maple from other programs

Contact Waterloo Maple Software for information. There is a *C*-callable version of Maple available. MATLAB is able to call Maple through its Symbolic Toolbox. Maple is also callable from IMSL subroutines.

References

[1]   American National Standards Institute/Institute of Electrical and Electronic Engineers: *IEEE Standard for Binary Floating-Point Arithmetic,* ANSI/IEEE Std 754-1985, New York, 1985.

[2]   Uri M. Ascher, Robert M. M. Mattheij, and Robert D. Russell, *Numerical Solution of Boundary Value Problems for Ordinary Differential Equations,* Prentice-Hall, 1988.

[3]   E. J. Barbeau, *Polynomials,* Springer-Verlag Problem Books in Mathematics, 1989.

[4]   Carl M. Bender and Steven A. Orszag, *Advanced Mathematical Methods for Scientists and Engineers,* McGraw-Hill, 1978.

[5]   Wolf-Jürgen Beyn, "Numerical Methods for Dynamical Systems," in *Advances in Numerical Analysis,* Will Light, ed., Oxford Science Publications, 1991, *pp.* 175–236.

[6]   William E. Boyce and Richard C. DiPrima, *Elementary Differential Equations and Boundary Value Problems,* 2nd. ed., Wiley, 1969, *pp.* 423–429.

[7]   Michael W. Chamberlain, "Heart to Bell," College Mathematics Journal 25, No. 1, January 1994, *p.* 34.

[8]   Bruce W. Char, Keith O. Geddes, Gaston H. Gonnet, Benton L. Leong, Michael B. Monagan, and Stephen M. Watt, *First Leaves: A Tutorial Introduction to Maple V,* Springer-Verlag, 1992.

[9]   Bruce W. Char, Keith O. Geddes, Gaston H. Gonnet, Benton L. Leong, Michael B. Monagan, and Stephen M. Watt, *The Maple V Language Reference Manual,* Springer-Verlag, 1991.

[10]   Bruce W. Char, Keith O. Geddes, Gaston H. Gonnet, Benton L. Leong, Michael B. Monagan, and Stephen M. Watt, *The Maple V Library Reference Manual,* Springer-Verlag, 1991.

[11]   Robert M. Corless, "Continued Fractions and Chaos," Amer. Math. Monthly 99, vol. 3, 1992, *pp.* 203–215.

[12] Robert M. Corless, *Symbolic Recipes*, vols. 1 and 2, Springer-Verlag, 1994.

[13] Robert M. Corless and Khaled El-Sawy, "Solution of Banded Linear Systems in Maple using *LU* Factorization," from *Proceedings of the Maple Summer Workshop and Symposium, Troy, New York, Aug. 1994*, Robert J. Lopez, ed., Birkhäuser, pp. 219–227.

[14] Robert M. Corless, Gaston H. Gonnet, D. E. G. Hare, and David J. Jeffrey, "Lambert's *W* Function in Maple," Maple Technical Newsletter 9, Spring 1993, *pp.* 12–22.

[15] Robert M. Corless, Gaston H. Gonnet, D. E. G. Hare, David J. Jeffrey, and Donald E. Knuth, "The Lambert *W* Function," *to appear.*

[16] Robert M. Corless and D. J. Jeffrey, "Well, It isn't Quite that Simple...," SIGSAM Bulletin 26, vol. 3, August 1992, *pp.* 2–6.

[17] Robert M. Corless and Michael B. Monagan, "Simplification and the assume Facility," Maple Technical Newsletter 1 no. 1, June 1994, *pp.* 26–31.

[18] W. H. Enright, 1989, "A New Error-Control for Initial Value Solvers," Appl. Math. Comput. 31 *pp.* 288-301.

[19] Temple H. Fay, "The Butterfly Curve," Amer. Math. Monthly 96, No. 5, May 1989, *pp.* 442–443.

[20] Walter Gander and Jiří Hřebíček, *Solving Problems in Scientific Computing Using Maple and MATLAB*, Springer-Verlag, 1993.

[21] Keith O. Geddes, Stephen R. Czapor, and George Labahn, *Algorithms for Computer Algebra*, Kluwer, 1992.

[22] G. Golub and C. van Loan, *Matrix Computations*, Johns Hopkins, 1983.

[23] G. H. Gonnet and D. Gruntz, "Limit Computation in Computer Algebra," Technical Report 187, Department for Computer Science, ETH Zürich, 1992.

[24] D. Gruntz, "A New Algorithm for Computing Asymptotic Series," Proc. ISSAC '93, Kiev, Ukraine, *pp.* 239–244.

[25] Xavier Gourdon and Bruno Salvy, "Computing One Million Digits of $\sqrt{2}$," Maple Technical Newsletter 10, Fall 1993, *pp.* 66–71.

[26] E. Hairer, S. P. Norsett, and G. Wanner, *Solving Ordinary Differential Equations I*, Springer series in computational mathematics 8, Springer-Verlag, Berlin, 1987.

[27] André Heck, *Introduction to Maple*, Springer-Verlag, 1993.

[28] Peter Henrici, *Applied and Computational Complex Analysis* vol. I, Wiley-Interscience, 1977.

[29] Peter Henrici, *Applied and Computational Complex Analysis* vol. II, Wiley-Interscience, 1977.

[30] T. E. Hull, W. H. Enright, B. M. Fellen, and A. E. Sedgwick, "Comparing Numerical Methods for Ordinary Differential Equations," SIAM J. Numer. Anal., **9**, 1972, *pp.* 603–637.

[31] D. J. Jeffrey, "The Importance of Being Continuous," Math. Magazine, **67** Oct 1994, *pp.* 294–300.

[32] Donald E. Knuth and Boris Pittel, "A Recurrence Related to Trees," Proc. Amer. Math. Soc., **105**, no. 2, Feb. 1989, *pp.* 335–349.

[33] John McCleary, "How Not to Prove Fermat's Last Theorem," Amer. Math. Monthly **96**, No. 5, May 1989, *pp.* 410–420.

[34] M. B. Monagan, *Programming in Maple,* available by anonymous ftp to the Maple Share Library.

[35] Michael B. Monagan and Walter M. Neuenschwander, "GRADIENT: Algorithmic differentiation in Maple," Proc. ISSAC '93, Kiev, Ukraine, *pp.* 68–76.

[36] Alexander Morgan, *Solving Polynomial Systems Using Continuation for Engineering and Scientific Problems*, Prentice-Hall, 1987.

[37] Ivan Niven, *Maxima and Minima without Calculus*, Dolciani Mathematical Expositions No. 6, Mathematical Association of America, 1981.

[38] T. S. Parker and L. O. Chua, *Practical Numerical Algorithms for Chaotic Systems*, Springer-Verlag, 1989.

[39] Darren Redfern, *The Maple Handbook*, Springer-Verlag, 1993.

[40] Theodore J. Rivlin, *Chebyshev Polynomials*, 2nd. ed., Wiley-Interscience, 1990.

[41] Bruno Salvy, "Efficient Programming in Maple: A Case Study," SIGSAM Bulletin **27**, vol. 2, April 1993, *pp.* 1–12.

[42] Helen Skala, "Contour Maps—A Visual Experience," The College Mathematics Journal **22**, No. 3, May 1991, *pp.* 241–244.

[43] Ernst Joachim Weniger, *Nonlinear Sequence Transformations for the Acceleration of Convergence and the Summation of Divergent Series*, Computer Physics Reports **10**, No. 5, North-Holland, December 1989.

[44] J. H. Wilkinson, "The Perfidious Polynomial," in *M.A.A. Studies in Numerical Analysis*, Gene Golub, ed. 1984, *pp.* 1–28.

[45] E. M. Wright, "The Linear Difference-Differential Equation with Constant Coefficients," Proc. Royal Soc. Edinburgh, **LXII** A, 1947, *pp.* 387–393.

Index